The Ergonomics Kit

FOR GENERAL INDUSTRY

SECOND EDITION

The Ergonomics Kit

FOR GENERAL INDUSTRY

SECOND EDITION

Dan MacLeod

CRC Press
Taylor & Francis Group
Boca Raton London New York

CRC Press is an imprint of the
Taylor & Francis Group, an **informa** business

A TAYLOR & FRANCIS BOOK

First published in 2006 by Taylor & Francis Group

Published 2019 by CRC Press
Taylor & Francis Group
6000 Broken Sound Parkway NW, Suite 300
Boca Raton, FL 33487-2742

© 2006 by Taylor & Francis Group, LLC
CRC Press is an imprint of Taylor & Francis Group, an Informa business

First issued in paperback 2019

No claim to original U.S. Government works

ISBN-13: 978-0-367-45382-4 (pbk)
ISBN-13: 978-0-8493-7029-8 (hbk)

**Visit the Taylor & Francis Web site at
http://www.taylorandfrancis.com**

**and the CRC Press Web site at
http://www.crcpress.com**

Library of Congress Card Number 2005033742

Library of Congress Cataloging-in-Publication Data

MacLeod, Dan.
    The ergonomics kit for general industry / Dan MacLeod.-- 2nd ed.
      p. cm.
    Includes bibliographical references and index.
    ISBN 0-8493-7029-9 (9780849370298 : alk. paper)
    1. Human engineering. I. Title.

T59.7.M32 2006
620.8'2--dc22   2005033742

In memory of Ardelle,
and as always,
for Bertrande

Preface

Since the publication of the First Edition of *The Ergonomics Kit for General Industry*, the field has continued to grow in popularity. There is increasing hard evidence from a variety of industries that well-developed ergonomics programs can reduce costs and provide considerable value to business. Comprehensive programs have been in place in many companies for a decade, and data has begun to be generated. Despite the controversy of the U.S. Occupational Safety and Health Administration (OSHA) regulation — which was rescinded by an Act of Congress — ergonomics is becoming firmly rooted in day-to-day business in the workplace.

The biggest financial value of workplace ergonomics has been from protecting valuable human resources by reducing costly musculoskeletal injuries. Most companies that have applied the ergonomics process systematically have enjoyed substantial reductions in workers' compensation costs.

Additionally, workplace ergonomics has increasingly shown its value in improving efficiency. In many companies, the ergonomics process is the only activity that is causing a task-by-task review of all operations with the explicit goal of finding improved methods. Consequently, employers are finding better ways of performing their operations. Similarly, many experts are discovering that the advanced methods of industrial work (whether in manufacturing or service industries) can only be optimized when the concepts of ergonomics are understood and applied.

However, to achieve good results employers need to incorporate an effective *process*. There are still too many cases where ergonomics efforts have failed because of inappropriate organizational systems. For example, some corporate programs are more bureaucratic than even the worst government agency. More commonly, task analysis worksheets and scoring systems are often too complicated to be much good. As a final example, there are still senior managers who do not understand what ergonomics is all about and the potential financial benefits it can provide. I hope the materials in this kit will help remedy these problems.

New Materials

The First Edition of *The Ergonomics Kit for General Industry* provided guidance on the fundamentals of setting up a program. This edition amplifies that advice with additional material on the actual process of solving problems. I have also added a number of success stories from my personal experience that serve to illustrate how the process works under various circumstances and organizational cultures. In each case, the emphasis is on simple, low-tech, and low-cost approaches.

One of the most frequently asked questions in the past decade has been "How do you create change?" Consequently a new section has been added to provide guidance.

The electronic versions of forms, worksheets, and PowerPoint presentations have proven to be popular with users of the First Edition. This edition continues this valuable support by making the electronic files available on the internet, where they are easier to update.

Everything in this book is still founded on practical experience. Nothing is based on what might "sound good" or be theoretically important. It is all rooted on what has actually succeeded on the workplace floor.

International Application

Sweden still leads the world in the development and application of the workplace ergonomics process, followed closely by the U.S. and Canada. The rest of the advanced industrial world from Singapore to Mexico to Spain is making great strides. The new industrial giants like China and India will undoubtedly follow soon, and with great vigor. Readers from these countries should find the ideas in this book as applicable in their homelands as in the U.S., where the materials were originally developed.

The book contains obvious reference to the U.S. and its legal and cultural background. Nonetheless, the process and concepts apply in any country. Often, the only differences between countries are the names of government agencies and their versions of recordkeeping forms. To be sure, there can also be a change in focus to some degree, as for example, the emphasis in countries like Singapore and Britain (at the time of this printing) on manual material handling or Sweden's emphasis on employee decision-making. But all in all, these materials should be helpful regardless of geography. The references to U.S. particulars should be easily understood and the concepts transferable.

Some Simplified Spelling

There is a great need to make English spelling more ergonomic and I have made a modest effort along these lines by using *thru*, *tho*, and *altho*. In the U.S., we use these simpler versions informally, but we seem to feel that we're doing something wrong in more formal settings.

In fact, these spellings were accepted by the 1898 convention of the National Education Association (NEA) as part of a mostly successful effort at that time to simplify several hundred spellings. Better spellings became common in the U.S. for words such as *draft* (instead of *draught*) and *catalog* (vs. *catalogue*). But for some reason, the words *thru*, *tho*, and *altho* never became common in formal writing as they were intended.

I would encourage all readers to adopt this practice in formal communication. These simplifications would represent a few more small steps along the way to a better system of spelling.

Program vs. Process

In recent years there has been a lot of commotion about *process* and its superiority over a *program*. This emphasis involves a valid point, that is (in this context), the term *process* highlights the importance of changing core practices, which is contrasted with *program,* which is considered to be an attempt to change surface behaviors without touching the underlying system. To some extent the term *program* has even become pejorative, indicating a sort of lip service to real change.

But this emphasis should not be taken too far. In many ways a *process* means more or less the same thing as *program.* Indeed, in this book the terms are sometimes used interchangeably.

On yet a different level, there are times when these words actually mean different things. There are occasions in this book — and these cases should be self-evident to the reader — that *program* is intended to mean the organizational structure that is established, such as committees and training classes, and *process* is intended to mean the sequence of events that bring about an action. For example, you would establish a *program* that is intended to change the *process* of handling problems in the workplace on a day-to-day basis.

Format

Because of the worksheets and checklists that are included, it was logical to keep the page size of the book the same as those materials. However, typewritten text across the full width of a page this size is difficult to read; columns of text are much easier to read.

Consequently, I adopted a style of subheads or illustrations along the left-hand column and text in a right-hand column. This format should help you easily find items that you need, plus make it easier to read. The extra white space also provides a place for you to jot down notes and ideas on how to adapt generic points to your facility.

Finally, for certain background comments like this, I have used a center column format. In total, I hope this provides a clean and friendly organization of the material.

Companion Book — *The Rules of Work*

This book is a companion to *The Rules of Work*, written by the same author and also from Taylor and Francis. Whereas *The Ergonomics Kit for General Industry* describes "how to set up the process," *The Rules of Work* describes "what is ergonomics," with many examples of how to solve specific problems and an introduction to common measuring techniques. Both books should be valuable in any workplace.

Acknowledgments

Everything in these materials is based on what I have learned working on-site in countless workplaces and with various organizations.

Thirty years ago, a large number of people from the United Auto Worker's union (UAW) helped me understand practicality in the workplace, learning to separate realistic guidance from its academic background. Carl Carlson, Dick Marco, Oscar Pascal, and Art Shy in particular were great teachers. The participants in the classes I led at the UAW's school in Black Lake, Michigan, provided a good testing ground for trying to learn to say things simply and clearly. I still follow the writing style that I learned to use for UAW safety and health publications.

Jim Morris helped me understand things from a business perspective. Devin Scott of the American Meat Institute (AMI) taught me a great deal about how trade associations work. Almost everyone I met in the meatpacking industry in the 1980s made me think in down-to-earth terms. A few years ago, I spent more time in a U.S. Postal Service processing and distribution center than in any other single workplace, continuing my education there from my Ergonomics Team leaders: Bobby LaDuke, a postal clerk; Mike Scheuer, a mailhandler; and Bill McCollough, a letter carrier.

The continuing stream of dedicated corporate staff that I have worked with has always provided me with insights and inspiration: Joe Pallansch, Anita Morris, Don Hirsch, Ron Allen, Alan Peterson, Jim Edwards, Dennis Taylor, and many others.

I owe much to many young professional ergonomists with whom I have worked, including the following very bright and helpful people: Eric Kennedy, Wayne Adams, Elizabeth Damann, Rob Nerhood II, Greg Worrell, Nancy Larson, Dan Cimmino, and Mark Heidebrecht. Professional colleagues Al McCarty, Jerry Combs, Bill Boyd, Mike Melnick, Rick Johnson, and Rick Pollack have always been informative and helpful. Cindy Roth is in a class by herself in terms of understanding how organizations work and how to work with them.

Various artists have provided the illustrations in this book: Ethan Greenbaum, Mark Watson, Mary Noyes, and Tom Nynas. Bertrande MacLeod added her invaluable editing talents.

To all of these people, I owe my thanks and appreciation.

About the Author

Dan MacLeod is one of the leading practitioners of workplace ergonomics, widely recognized for his down-to-earth approach and his emphasis on low-cost, low-tech job improvements. He served as a consultant to employers, trade associations, and unions for over 30 years, and has developed innovative corporate programs that have resulted in thousands of ergonomic improvements and savings of millions of dollars. In all, he has conducted evaluations in over 1200 workplaces in his lifetime, assessing thousands of individual tasks.

He was instrumental in the late 1970s in recognizing the widespread impact of work-related musculoskeletal disorders. In 1982, he authored the first lay language training materials in North America on workplace ergonomics, a work that was later republished in four foreign languages. He has comprehensive experience in multiple industries, ranging from the office environment and hospitals to underground mining and steel mills. He has worked with small organizations as well as some of the largest corporations in the U.S.

His work has helped change entire industries. In the late 1970s, he pioneered the first workplace ergonomics programs in car assembly operations. In the 1980s, he was heavily involved in the meatpacking industry, where he represented the industry in working with OSHA to develop the meatpacking guidelines.

He has worked and traveled widely in Europe, Central America, and Asia. He speaks several languages with varying degrees of fluency: Swedish, French, German, and Spanish, plus rudimentary Chinese.

Dan is the author of other full-length books on practical ergonomics, plus a wide variety of training manuals, booklets, and videotapes. He is a Certified Professional Ergonomist (CPE) and holds master's degrees in both occupational health and industrial relations.

He works as an independent consultant. For more information, see www.danmacleod.com.

Contents

Electronic Versions

Many materials from this book are available electronically:
- Worksheets, checklists, audit forms, and training handouts
- A PowerPoint® training presentation

You may access these files on the downloads page of the publishers website at www.crcpress.com or by using the following direct link: www.crcpress.com/e_products/downloads/download.asp?cat_no=7029

Copyright Notice

Introduction

So now you're supposed to do ergonomics!?

If you're like a lot of people, you've just been volunteered to take on responsibilities for ergonomics in your workplace (and you hardly know how to spell the word). Maybe you're even the coordinator of the effort. Undoubtedly you have to add this responsibility to many others that you already have.

But fear not. This kit gives you ample support and guidance.

Don't be overwhelmed by all the material in the following pages. You don't need to read or use *every*thing. Many sections of the book may not apply to your situation at all. However — and this is a point of reassurance — if you do need help on a certain topic, it is likely that there is support material for you.

Moreover, the ergonomics process is flexible enough to support your other responsibilities, and indeed, may make many of these other duties easier. Finally, ergonomics can be fun and rewarding, so being active in the ergonomics effort in your workplace need not be a burden at all.

Five Ways to Get Started

1. Read the introduction to this book to begin learning the basics of what you need to look for and do.

2. Go around and look. The key is simply putting on your "ergonomics glasses," looking at jobs, and thinking of ways to make things better. You don't even need to use any of the checklists at first. Start getting some experience under your belt by making a few obvious changes. Even simple things can provide a good beginning.

3. Form a team — "TeamErgo" — either a site-wide group, or a subcommittee of your Safety Committee, or any other way to gain some strength in numbers.

4. Do the brainstorming exercises on getting started in the pages of Chapter 10 *Worksheets*.

5. Review injury and illness logs such as the OSHA 300 form. Identify the ones that appear to be related to over-exertion or repetitive motion.

Putting on Your Ergonomics Glasses

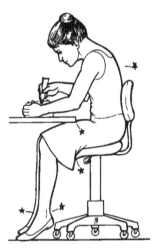

Much of the task before you simply amounts to "putting on your ergonomics glasses," that is, learning to look at routine activities from a new perspective.

While there is certainly much more to the field of ergonomics than the basic information provided in this kit, learning and then applying these concepts do not need to be forbidding tasks.

The Next Steps

6. Read or skim Chapter 4 *Case Examples* to get a feel for what some other successful organizations have done.

7. Review the tips and ideas described in the rest of the book that appear to apply to you.

8. Begin to write down your own process and plan based on Chapter 5, *Setting Up a Program*, plus using the worksheet, *20 Steps to Develop a Plan,* found in Chapter 10. Just get started. You can flesh out a full program later.

9. Keep looking. Nothing beats spending time looking at jobs, thinking about ways to make improvements, and implementing self-evident changes.

10. Get ready to have fun and gain personal satisfaction by making your workplace safer, better, and more efficient.

Multiple Benefits

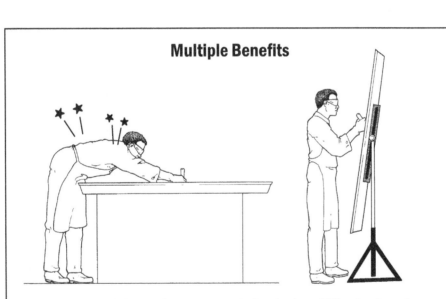

In the illustration above left, it is difficult for the employee to see and difficult for him to make accurate movements. Plus, it is fatiguing and painful on his back. It all adds up to making it hard to do the job well.

By using a tilter for the product, as shown at right, the setup enables greater accuracy, easier actions, more energy, and less pain. All of this results in being able to accomplish the task more efficiently with better quality and higher employee morale. The ergonomics process can help you find simply better ways of doing work.

What is Ergonomics?

Fit the task to the person

Probably the best phrase that describes the field is: "Fit the task to the person, rather than the person to the task." Whenever we set up a piece of equipment, we need to ask: "How does the human fit in?" When designing a tool or planning a task, we need to consider human strengths and limitations.

Make things user-friendly

The term user-friendly is identical with ergonomics. Anything that can be described as user-friendly can also be said to be ergonomic. Conversely, unfriendly items are not ergonomic.

User-friendly means that things are easy to understand and apply, that mistakes are reduced, and that the human is treated well in the process. The concepts apply to both physical issues as well as mental or cognitive ones.

Work smarter, not harder

A time-worn phrase that almost everyone seeks to do is "work smarter, not harder." Normally, how one actually goes about doing so is left unstated. But ergonomics remedies this by providing a method for finding smarter ways of performing manual work.

The rules of work

Finally, it is instructive to know that the term ergonomics was coined from the Greek words *ergon* (meaning "work") and *nomos* (meaning "rules"). So the literal meaning is "the rules of work," which has a nice ring to it. We all need to know the "rules" for optimizing work.

Formal definition

Ergonomics: the field of study that seeks to design tools, equipment, and tasks to optimize the interface between humans and systems.

This interface can be as simple as that between a human and a chair (such as the back rest, the cushioning, or the height) or a much more complex interrelationship between an employee and an entire production line.

Other terms often used for the field include:
- Man-Machine Systems
- Human-System Interface
- Human Factors Engineering

MSD Prevention

Ergonomics has gained recognition in industry for its value in preventing musculoskeletal disorders (MSDs), a class of injuries that basically amounts to "wear and tear" on the human body (and described in more detail in Chapter 3, *Understanding MSDs*). Consequently, there is a tendency to use the two terms interchangeably, which is *not* technically correct.

As we will see, there is certainly more to ergonomics than preventing MSDs and vice versa. As an example of this confusion, OSHA's "ergonomics" standard was a misnomer and really should have been called "MSD Prevention Standard."

Applications

The possible uses of ergonomics include just about every activity that people do both on and off the job.
- Manufacturing
- Service industry
- Office work
- Home chores and leisure activities
- Consumer products

Ergonomics provides a way of thinking about all of these tasks and how to make them more efficient and easier to do. In these materials we are going to focus on the workplace. But just remember that you can apply what you learn to almost any activity.

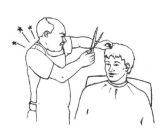

Service Jobs

Leisure

Home Chores

Perspectives

Before we go any farther, we need to adopt a few practical perspectives. The term "ergonomics" might sound hard and it has an undeserved reputation for involving expensive products. Furthermore, it sounds like a new field of study, maybe one that is a bit too far-fetched for the hard realities of modern business life. But it's not necessarily so.

It doesn't *have* to be hard

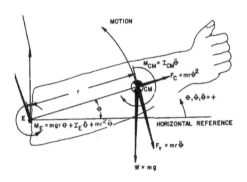

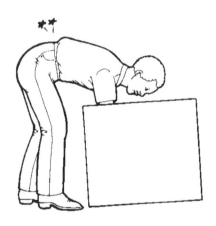

For almost every aspect of ergonomics, we can address the issues on two levels: (1) as a subject for sophisticated science and (2) as a matter of common sense. Both are equally important, but the latter takes more priority in most workplaces.

Sophisticated science

On every issue, we can apply the scientific method. Researchers can conduct rigorous studies, measure human attributes, and build mathematical models. There are topics such as "biomechanics" (studying the human body strictly as a mechanical system) and "anthropometry" (measurement of humans) that require special training and experience.

There is an increasing stock of measuring techniques that can be applied in the workplace. These methods have value, but some can be daunting for people who aren't specialists in the field.

Common sense

But it doesn't have to be that complicated. Much of ergonomics amounts to common sense once you start thinking about it. Ordinary people can provide many ideas about improving the tools, equipment, and tasks in the workplace, particularly once they receive training in basic principles.

Most everyone has done "ergonomics" all their lives, without using the word. We all have moved things around or used a tool to make a task easier for us.

There's an ergonomist in every one of us. We all have certain inclinations to modify our surroundings to fit us. Much of ergonomics can be common sense — once you put on your ergonomics glasses.

It doesn't *have* to be expensive

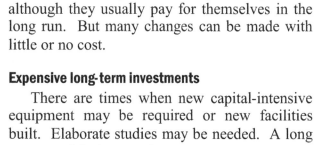

The expense of applying ergonomics can fall within a wide range. Some things are expensive, although they usually pay for themselves in the long run. But many changes can be made with little or no cost.

Expensive long-term investments

There are times when new capital-intensive equipment may be required or new facilities built. Elaborate studies may be needed. A long process of design may be necessary.

As an example, the workstation that has been studied more than any other using formal techniques of ergonomics is the airplane cockpit. Evaluating and improving the design of an airplane cockpit is not an inexpensive proposition, either to determine what types of changes might be made, or to implement them. However, it does have a clear payback of helping pilots fly safely.

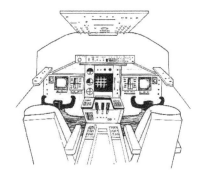

Similarly, large-scale industrial processes may take considerable resources to investigate and design properly. Some equipment indeed is more costly than the standard alternatives. The resulting improvements may still have a positive financial return but the changes may demand a large initial investment.

Low cost

But improvements do not always need to be expensive. It is often surprising what can be done for low cost by using a little imagination. Some companies estimate that there are 20 low-cost fixes for every one that takes some investment.

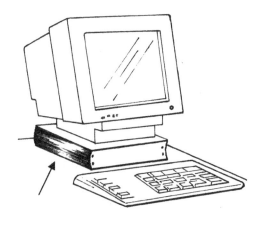

As an example of a quick fix, you can raise a computer monitor by placing it on top of a thick book. Awkward heights can often be made better by such simple measures. Indeed tasks can often be improved with a variety of low-cost, or even zero-cost, changes:
- Rearranging layouts
- Adjusting equipment properly
- Inexpensive rests and padding, etc.

More detailed examples of cost-free or low-cost improvements are provided in this book. As part of the day-to-day workplace ergonomics process, you should deliberately seek these opportunities.

It's not necessarily *new*

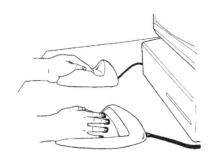

Sometimes it seems like ergonomics is all brand new, because we may have heard the term for the first time in recent years, or because we see unusual products available on the market.

Consequently, it is possible to get the impression that ergonomics is not quite ready for the here and now. Some people might even have the notion that it might be best to wait a while to see if ergonomics is just a passing fad.

But this isn't quite right; in many ways, ergonomics is nothing new. In a way, humans have been doing ergonomics for about 40,000 years, ever since the first human picked up a rock to use as a hammer or a stick as a pry bar.

A good example is a great ergonomic product invented in the 19th century — the long-handled scythe. Note the ergonomic features of the two-handed scythe compared to a one-handed sickle.

You can work upright, keeping the natural curves of the lower back; the grips move up and down to adjust for the farmer's height; the grips can be angled to keep the wrists in their neutral postures; and the very shape of the scythe takes advantage of the larger muscles and mechanical structure of the upper torso.

And the bottom line is well known. The ergonomic device is clearly easier on the human and much more productive than the non-ergonomic device. (And in a spirit of continuous improvement, the mechanical harvester later on was even better.)

The big difference between old and new

There is a difference, however, between old-fashioned and modern ergonomics. Old-fashioned ergonomics was haphazard, a process of trial and error thru the years. Modern ergonomics has the advantage of the scientific method and being systematic. Today we can measure, use analytic techniques, and refer to a growing database of knowledge.

The promise of modern ergonomics is that if we systematically evaluate every task that we do, we can systematically develop equivalents of the two-handed scythe. What now lies before us is to take this natural tendency to modify our surroundings for our benefit and turn it into a conscious approach to management and design.

Principles of Ergonomics

The following are 10 basic principles that summarize the field of physical ergonomics. There is some overlap to these principles, but taken as a whole, they represent a set of prescriptive statements that can help you learn what to look for and find smarter ways to work.

1. Work in neutral postures
2. Reduce excessive force
3. Keep everything in easy reach
4. Work at proper heights
5. Reduce excessive motions
6. Minimize fatigue and static load
7. Minimize pressure points
8. Provide clearance
9. Move, exercise, and stretch
10. Maintain a comfortable environment

Each of these principles is described in more detail in the following section, including a generic explanation and some examples. Note that there are thousands of good examples that could be shown, but these few should suffice to give you a good idea of the basic concept.

Also, note that while many of these principles may appear simple and self-evident, they are routinely unmet in the workplace. Furthermore, one should not underestimate the power of a few fundamental ideas applied systematically.

The Power of a Few Principles

A good example of the need to emphasize principles rather than specific applications is the debate about using wrist rests at computer workstations. In reality, the issue is not whether or not wrist rests should be provided to everyone; rather, the issue is the posture of the wrist. If the posture conforms to good principles, then a wrist rest is not needed. If the wrist is bent, a wrist rest is one of many ways to make the needed improvement. Understanding the underlying principles helps resolve the disputes.

Consequently, it is much more important for you to learn the underlying principles rather than the details of current prescriptions for specific problems.

- By learning the principles you will understand how to evaluate changes in technology and new products that will become available in the future.
- The principles will help you to evaluate any task, whether at home, in the office environment, or in general industry.
- The principles will remain the same, even when advancements in knowledge are made in the field of ergonomics

Low Tech

The approach of the following section is low tech and differs from the quantitative style of ergonomics that emphasizes the evaluation of workstations in terms of centimeters, degrees, and precise placement. Quantification can be important, but for practical purposes, this down-to-earth orientation is generally more accessible and useful, particularly for newcomers to the ergonomics process.

Caveats

It is important for you not to misinterpret these principles. It is not possible to meet all these conditions for every job and no claim is made here that the improvements are always feasible, either technologically or economically. These principles are design goals, not rigid requirements. The point is to keep them in mind as you evaluate jobs and take advantage of the opportunities that do arise when improvements are feasible.

Principle 1
Work in Neutral Postures

Working in awkward, contorted postures increases
fatigue and physical stress on the body, plus reduces strength
and dexterity, thereby making it more difficult to do a task.

Maintain the natural curve of the spine

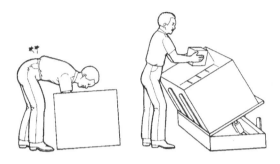

One of the most common things to look for
when evaluating jobs is the position of the back
— you don't want to work for sustained periods
with a bent or twisted back. If you see a situation
like this, then you typically want to start
thinking about ways to make changes so that the
person can work in an normal upright position.

In its natural position, the lower back has a
slight arch or "swayback." You need to maintain
this arch whether sitting or standing.

Keep the neck aligned

The neck is part of the spinal column and
thus is subject to the same requirements as the
lower back. Consequently, another issue to look
for is prolonged twisting or bending of the neck.
You put on your ergonomics glasses and observe
people who might be working in an awkward
neck posture. If you see anyone like this, then
you think of possible improvements.

Keep elbows in and shoulders relaxed

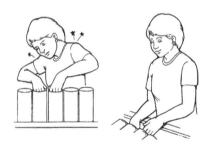

The optimal position for your arms is with
your elbows held comfortably at the sides of
your body. The shoulders should be relaxed and
not hunched. If you work with your elbows
winged out, it can add strain to the shoulders and
cause fatigue and discomfort. Plus, working
with the arms outstretched, like other awkward
postures, can reduce strength and dexterity and
thus interfere with people's ability to do their
jobs well.

Keep wrists in neutral

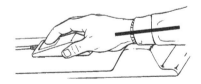

The hands should be kept in the same plane
as the forearm, which is to say that the wrists
should not be bent or twisted. There's more to
the issue of wrist posture than this (as there is to
the whole topic of working posture), but this
simple direction should be enough to get you
started.

A final remark: the message here is *not* that
you should *never* deviate from these postures.
The points apply to sustained work . . . and when
feasible.

Principle 2
Reduce Excessive Force

Excessive force can overload the muscles, creating fatigue and potential for injury. Furthermore, if you need to apply excessive force to perform a task, it can slow you down and interfere with your ability to perform the task well. Consequently, almost anything that you can do to minimize the exertion required for the task will make it easier and more efficient.

Loads on the back

Push/pull forces on the arm

Grasping and pinching force

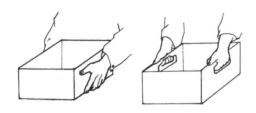

In addition to looking at the *posture* of the back, you need to evaluate the *load* on the back too, that is, if there is any lifting, pushing, and/or pulling. Likewise, in addition to evaluating the postures of the arms and wrists, you must consider the loads on these joints as well.

Then, if you notice excessive exertion, you should start thinking of ways to reduce the force. It is possible to measure many these forces to see if guidelines are exceeded, but usually it is better to immediately brainstorm possible alternatives. Maybe there is an inexpensive way to make it better.

There are countless ways to reduce force. In fact, finding ways to reduce force has been one of the hallmarks of human progress over past centuries. It is helpful to think in terms of basic strategies that can be used depending upon the situation:

- levers
- conveyors, slides, and skids
- counterbalances
- improved grip design
- changed methods
- using body position to best advantage
- fixtures and backstops
- tools and machines

What is important is to get people thinking about what they do and get into the habit of always searching for better ways. At this level, ergonomics is not technically difficult. The hard part is learning to see things that you normally walk right by and then challenging "the way we've always done it."

Principle 3
Keep Everything in Easy Reach

An easy way to make your work more user-friendly is to keep products, parts, and tools that are frequently needed within easy reach. Long reaches often cause you to twist, bend, and strain, which makes work more difficult and time-consuming.

Reach envelope

A good rule is to keep frequently used materials within the "reach envelope" of the arm. Moreover, things you use almost constantly should be within the reach envelopes of the forearms. Note that this envelope is a semi-circle, not the rectangle that we typically use for work surfaces.

The easiest way of applying the rule is simply to see if it is possible to rearrange the layout of tools and equipment. It is surprising how often we can move things closer. We easily become so used to reaching that we don't even notice it, even tho it may be causing problems.

In some cases it is helpful to reduce the size of a given work surface. As another alternative, if you need a lot of work surface but still have trouble with far reaches, it is possible to have a cutout made within the work surface rectangle.

Tilt

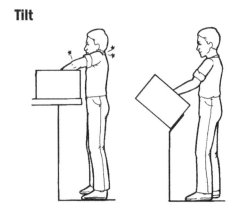

A common problem in many industries is reaching into or working out of containers. Possible improvements are:
- Propping up a container on one end
- Tilt tables or stands
- Lazy Susans
- Spring-loaded bins
- Adjustable-height stands
- Containers with drop-down sides
- Chutes and hoppers
- Smaller lot sizes and thus smaller containers

Design for shorter individuals

A useful rule of thumb is that reach distances should be established to accommodate smaller-statured people when possible. The idea is: If shorter people can reach, so can everyone else. (We will see a reverse rule for clearance in Principle 8).

Principle 4
Work at Proper Heights

A common workplace problem is a mismatch in heights between employees and the work that they are doing. This leads to poor postures and related fatigue, discomfort, and potential damage to soft tissue, plus, quite often, unnecessarily harder work.

Elbow height

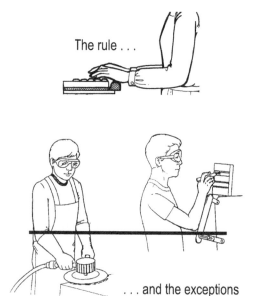

The rule . . .

. . . and the exceptions

Generally, work is best done at about elbow height, whether sitting or standing. This is true for computer keyboards plus other kinds of work such as manufacturing and assembly.

Note that it is the *work itself* that should be at elbow height, not the desk or work surface. For example, if unusually large products are being used, the height of the supporting surface should be adjusted accordingly. The issue is the height of the task being done, not the height of the work surface.

The nature of the work also affects the proper height. Heavier work, requiring upper body strength, should usually be lower than elbow height. Lighter work, such as precision work or inspection tasks, should be higher.

Consequently, you must take into account the nature of the task when designing proper heights. It is not always sufficient to look up a number in a book.

Adjustable heights

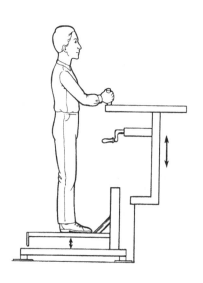

Since people vary in height, good design usually involves providing some sort of height adjustment. There are a variety of ways to meet this need.

When possible, the best approach is to adjust the height of the workstation itself, either permanently for an individual or as needed when several people use the same workstation.

An alternative is to provide a platform for shorter people. This approach is usually second best, since it creates more congestion, requires stepping up and down, and/or creates a potential trip and fall hazard. However, a standing platform is sometimes the only realistic option.

Providing for adjustable height for materials being worked with is also important. Methods like powered pallet lifts and spring-loaded surfaces or dispensers have been successful.

Principle 5
Reduce Excessive Motions

The number of motions required to do a task can have a profound impact on both efficiency and wear and tear on the body. Excessive motions can create injury to sensitive tissue and joints, as well as waste time. Whenever feasible, reduce unnecessary motions.

Design for motion efficiency

A good way to reduce motions is to improve layouts and organization. Materials should be presented in the correct orientation as close as possible to the point of use. Smooth flow of materials at correct heights and in correct sequence also reduces wasted motions.

Many of these ideas amount to old-fashioned methods engineering, ideas that have perhaps been neglected in an era of high technology. Nonetheless, consciously striving for motion efficiency is a strategy that can be readily used in many workplace ergonomics activities.

A good strategy in many instances is to slide items that must be handled repetitively, rather than pick them up one at a time to place in their locations. Note that although there are still motions involved when sliding items, the total number of motions is usually reduced.

Individual work methods are also part of the solution. It is not uncommon to see two individuals doing the same task, one smoothly with few motions and the other inefficiently.

Let the tool do the work

One of the best ways to reduce repetitions is to allow machines and tools to do the movement for you. Machines and power tools are good at performing repetitive tasks endlessly — so let them do the work.

There is nothing particularly remarkable about this concept; this is something we all do every day. Yet, it is worth mentioning because there is often confusion about how to reduce repetitive motions, and sometimes the obvious gets overlooked.

Motion-saving mechanisms

There are numerous mechanical techniques that can be use to reduce repetitive motions, such as gearing or rack-and-pinion, where one motion yields multiple turns. Ratchets and Yankee-drill mechanisms also reduce motions.

Principle 6
Minimize Fatigue and Static Load

Overloading physical and mental capabilities can lead to lost production, poor quality, accidents, and wear-and-tear injuries. Fatigued muscles slow you down plus are more prone to injury.

Metabolic load

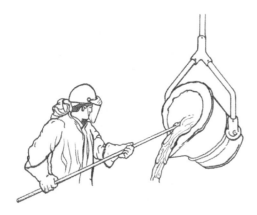

Fatigue can obviously result from heavy, exhausting activity, where you sweat and burn calories. This activity is known as *metabolic load* (from "high metabolism") and has been largely eliminated in the 20th Century as a result of machines, mechanical equipment, and robots.

Other ways to reduce this type of fatigue are to:
- Spread peak loads over more time
- Take frequent, short rest breaks
- Rotate with less demanding tasks
- Add staffing

Static load

Holding the same position for a period of time, known as *static load*, can cause pain and fatigue. A good example of static load that everyone has experienced is writer's cramp. After you hold onto a pencil for a while your muscles tire and begin to hurt. You do not need to hold it with a death grip for this to happen — just holding it loosely for a long time can make your hand hurt.

Pain and discomfort from static load is noticeable within minutes. Furthermore, continuous static loading of the muscles over months and years can contribute to long-term damage to tissue in the hand and wrist.

Offload the muscles

One of the best ways to reduce static load is to find some mechanical way of supporting the load other than with your muscles. Often, these devices can be quite simple and inexpensive.
- Arm rests to support outstretched arms
- Fixtures, straps, or hooks to hold an item
- Shelves or rails on which to support a load

Other approaches include: reducing force, improving working posture, and providing for rest breaks or opportunities to move. Examples of movement include sit/stand workstations or footrests (like the brass rail in a tavern).

Principle 7
Minimize Pressure Points

Direct pressure or "contact stress" is a common issue in many workstations. In addition to being uncomfortable, it can inhibit nerve function and blood flow.

Provide padding for hand grips

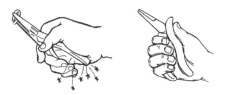

Provide padding for forearms

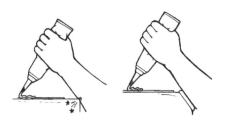

Provide cushioning for feet

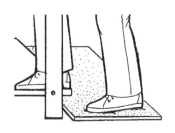

Chair cushioning

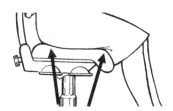

An example of pressure points that almost everyone has experienced is gripping hard onto a pair of pliers. The edges of the metal grips dig into your hand and can create considerable pain and discomfort.

Changing the shape, contour, size, and covering of tool handles can distribute the pressure more evenly over of the palm.

A similar problem is leaning your forearms against sharp or hard edges for support. The goal is to distribute the pressure over a larger surface area of the body. Improvements include:

- Padding the edge
- Rounding the edge
- Providing arm rests
- Redesigning the task
- Changing layout to avoid leaning

Standing for long periods of time on hard surfaces (especially concrete floors) can damage tissue in the heels, contribute to other leg disorders, and increase fatigue. Options for making improvements include:

- Anti-fatigue mats
- Fiberglass grating on mounts (for chemical facilities or sanitized food processing plants where mats often cannot be used)
- Cushioned insoles

The epitome of pressure points is sitting on a metal folding chair for a long time. Clearly, good cushioning is important for chairs.

Proper seat height also affects the pressure points. If the seat is too high, the legs dangle, creating pressure behind the knee. If the seat is too low, the weight of the body presses on the buttocks.

Principle 8
Provide Clearance

It is important to have both adequate workspace and easy access to everything that is needed, with no barriers in the way. Lack of clearance can create bumping hazards or force you to work in contorted postures. It can increase long reaches, especially if there is inadequate space for knees or feet.

Design for tall people

In general, the goal is to make sure that tall people have enough clearance, that is, room for the head, knees, elbows, and feet. If tall people can fit, then so can everyone else. To improve access:

- Reorganize equipment, shelves, etc.
- Increase the size of openings
- Eliminate obstructions between the person and the items needed to accomplish the task

A common problem is lack of knee or thigh clearance on desks, workbenches, or other types of equipment where people sit. Fixes include:
- Thin surfaces, with no hindering drawers
- Removal of obstacles

Maintainability

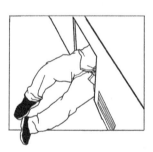

Probably the single biggest problem that maintenance personnel encounter in their tasks is lack of clearance. Many activities would be easy to perform, if you could only reach an item and work on it with easy access. Unfortunately, too often the items to be maintained are buried deep within the equipment. The remedy is designing with easy access in mind — quick disconnects, removable panels, improved configuration, and relocation of frequently accessed equipment.

Provide visual access

A similar issue is visual access. Visual access is the ability to see what you are doing or to see dials and displays.

A common issue is inability to see when moving a cart or lift truck. Equally common are machines where various gauges are distant from the operator's position. General workstations can also suffer from the same problem, much of which can be improved by removing barriers and changing layouts to provide better line of sight.

Principle 9
Move, Exercise, and Stretch

The human body needs to exercise and stretch. You should not conclude after reading all the preceding information about reducing repetition, force, and awkward postures, that you are best off just lying around and pushing buttons.

To be healthy, the human body needs activity. You need to stretch to the full range of motion for each of your joints occasionally throughout the day. Your heart rate needs to rise for a period of time every day. Your muscles need to be loaded on occasion.

Unfortunately, most jobs don't promote these activities. And where there is movement or exertion, it's often too much of the wrong kind.

Warm-ups and "energy breaks"

For strenuous tasks, it is helpful to warm up beforehand, just like any professional athlete would. It is not uncommon these days for employees like order pickers in warehouses to start the workday with group exercises.

For sedentary tasks, it is good to stop and stretch from time to time and even do some aerobic activity. Some companies are building these types of warm-ups and "energy breaks" into daily work life. In these situations, fitness instructors come into the work area, put on music, and lead the employees in a brief routine. In those companies that have been providing these energy breaks for some time, the group exercise has become ingrained in the culture and no one thinks twice about participating.

Allow for alternate postures

It is good to change and move. There is no one "correct" posture that is best for an entire workday. Adjustable equipment can facilitate such movement, but even without it you can often still change positions. If you have adjustable equipment, take advantage of it.

An increasingly common approach is the "sit-stand" workstation. Sometimes, the whole worksurface can be designed to move up and down. Other times it is possible to design the workstation for a standing posture, and then use a tall stool to sit on as needed.

People who routinely sit should get in the habit of changing the position of their chairs: raise it, lower it, tilt it forward, lean back, etc.

Principle 10
Maintain a Comfortable Environment

Humans often do not perform well in less-than-ideal environments. Excessive heat and humidity slow us down; excessive cold hinders effective work. Toxic chemicals can damage our health; vibration can injure sensitive tissue.

This principle is more or less a catch-all category in ergonomics. Some issues are normally addressed by other fields, such as industrial hygiene for toxic chemicals. But other issues, such as lighting, have been raised as part of the growth of workplace ergonomics.

Good lighting

The quantity and quality of light at the workstation can either serve to enhance or obscure the details of the work. Common problems include:

- Glare that shines in your eyes
- Shadows that hide details
- Poor contrast between your work and the background

One good way to improve lighting is use of task lights, rather than trying to provide all lighting from the ceiling. There are a number of important advantages of task lighting, including better control of glare, shadows, and brightness at each person's workstation.

Other ways to make improvements include:

- Diffusers or shields to minimize glare
- Better placement of lights or equipment

Isolate vibration

Working with vibrating tools or equipment can potentially cause carpal tunnel syndrome, plus other types of disorders such as Reynaud's phenomenon ("white finger"). Vibration can be reduced with:

- Better tool design
- Preventative maintenance
- Dampened tool grips
- Vibration-dampening gloves

Temperature extremes

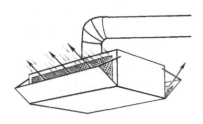

Sometimes the source of extreme heat and cold is inherent in the work (such as outdoor work, heat around furnaces, and the cold in meat and poultry plants) and little can be done about root causes. However, some steps can be taken to avoid some specific problems, such as using deflectors to keep cold air from blowing directly onto people.

Other Subfields

This kit deals almost exclusively with *physical* issues like exertion, heights, and repetitive motions. But there are other, equally important spheres within the overall field of ergonomics.

The most important other branch is *cognitive* ergonomics (sometimes called human factors engineering). Historically, cognitive ergonomics has been the larger and more advanced branch compared to physical ergonomics.

Cognitive ergonomics

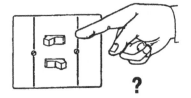

Cognitive ergonomics addresses how we conceive information, process it mentally, and decide on correct responses. By designing displays and controls — and in fact every type of information that we handle mentally — to take into account human perceptions and expectations, it is possible to reduce errors and improve performance.

A good example is the common light switch. Most North Americans would expect that if you want to use the switch illustrated at the left to turn lights *on,* you would flip *up* the switch. The design conforms to our expectations.

However, when confronted with a nonstandard design, such as that in the lower illustration, we get confused. The design is void of any feature that helps us decide which way to flip the switch. Moreover, people in other countries may react differently or have completely different types of light switches, which is part of the issue.

Broad scope — The application for cognitive ergonomics in designing things so that they are not confusing is incredibly broad. Examples include: television remote controls, road signs, written directions, bureaucratic forms, and warning signs.

Design principles — As with physical ergonomics, there are a number of general principles that can help guide us when we design or build things. A few examples are:

- Standardize — Make sure that similar devices work the same way or that color coding of wires and pipes is consistent.
- Use "stereotypes" — Incorporate common conventions, like red for warning or stop, or turning a dial to the right to increase speed or power.
- Use patterns and images — Take advantage of our visual capability to recognize shapes and patterns very easily (and our tendency *not to read* signs).

Work organization

A final general area of ergonomics has to do with the underlying design of work. Examples include:

Task allocation — How should tasks be divided and assigned to accomplish goals? Is it better to have many people equally capable of doing many tasks? Or is it better to have a narrow division of labor, so that individuals can be extremely highly qualified at specific tasks?

Assembly line or work cells — Should the technology and equipment of the workplace be designed so that tasks are narrowly defined? Or should the physical layout promote team activities?

Shift work — Should there be more than one shift in a given workplace? And, if so, should employees be assigned to just one shift (thus prohibiting some people from enjoying normal life), or should they be rotated between shifts every couple of weeks (thus forcing everyone to disrupt their biological time clocks)?

Reward systems — How should people be compensated for their activities? What are the actions that are rewarded? Should people be compensated for how much they put into a task (hours and effort), or how much they put out (quality and quantity of product)?

Other fields of study have obviously addressed these issues in great detail, but ergonomics provides a certain perspective. For example, management science and organizational psychology address many of these areas, but they do not always capture the technical side. Conversely, engineering addresses much of this subject, but it does not always capture the human side. The combination of the two is what interests ergonomists — the interface between humans and systems.

You're Off and Running!

Armed with just these simple ideas and principles, you can make many improvements in the tasks that people perform in your workplace. As you continue to learn more details on each of the principles, you will continue to increase your ability to identify issues and ideas for improvement.

Part I – Why?

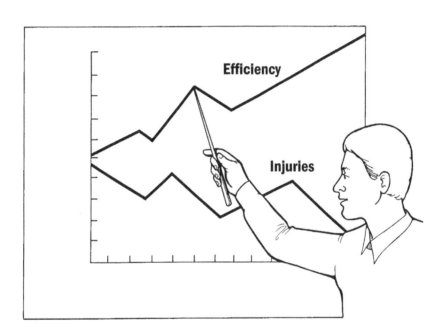

Electronic Summary

A summary of Part I, entitled 25 Ways Ergonomics Can Save You Money, *is available electronically. This enables you to print it and provide it to managers and supervisors as part of your effort to gain commitment and involvement. See the Table of Contents.*

Chapter 1
Good Ergonomics is Good Economics

The following story is an excellent illustration of the business value of the workplace ergonomics process. The bottom line is that an employee, after attending a class in ergonomics, came up an idea that was cost-free, eliminated a painful activity, and yielded 90% reduction in time needed to complete the job.

Vehicle mechanic

The company was a distribution operation with a large truck repair facility. Every year, some of the older delivery vans were cleaned up, repaired, and sold. The job involved removing the decals that covered much of the surface of the vans. The procedure — *"The way we've always done it"* — was to take a small razor blade tool and start scraping. Normally it would take a day or two to scrape a whole van. The task was unpopular and was traditionally assigned to the person with the lowest seniority.

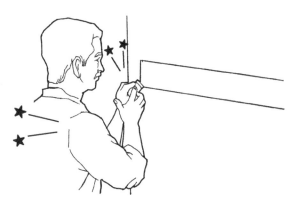

Before — Scraping the decals with a razor tool.

Painful shoulder

One morning, an employee was told to start scraping off the decals from a series of vans. By noon, his shoulder and hand were aching. By quitting time, he had almost completed one van and he was in pain. He looked at the work order and he discovered that he had 19 more vans to clean — he was going to be scraping all month.

Then he remembered our class: "The ergonomics guy said if we had a problem we should tell someone." So he sought out the Safety Committee Chairman, who happened to be the union steward for the facility and someone who I had worked with previously on other issues.

The Safety Committee Chairman, well familiar with the task, said, "Well, the ergonomics guy said that the most important thing to do is *think*." So the two walked to the scraping area and started brainstorming. After a time, one suggested using the power wash (that they normally used to clean the truck) to soak

Brainstorming

the decal and maybe to loosen it. They discovered that the hot water heated up the aluminum skin of the van and made it easier to remove the decals. So they turned up the heat of the water, played it on the van for a few minutes, then were astonished to find that the decals easily peeled off.

The time needed to remove the decals fell from one or two days per van to one or two hours. The solution was free, since the power wash was already on hand.

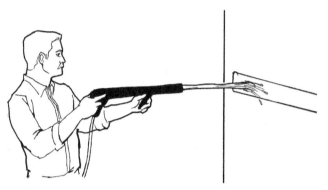

After — The power wash heats the surface of the van.

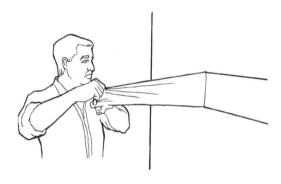

The decals peel off.

Thinking is good

In my view, this story represents the ideal of ergonomics success stories: it solved a problem that was hurting people; it resulted in dramatically greater efficiency; the employees themselves came up with the idea; and it was *free*. All that was needed establishing a workplace process that got people in the habit of challenging "the way we've always done it."

How to Understand Ergonomics in a Way That Makes Good Business Sense

Many people have gotten the false impression that ergonomics is a burden on business. Part of the explanation is historical — ergonomics gained visibility in much of business and industry because of regulation and litigation. The first exposure that many managers had to the field was media accounts of the multimillion dollar fines levied by OSHA against companies for problems related to poor ergonomics.

The reality is that ergonomics is good for business. To be sure, no claim is made here that *all* problems have an ergonomics solution or that in *every* instance the benefits will pay for the costs. Ergonomics is a problem-solving tool, one among many that organizations have. And like every other tool, it has its applications and its shortcomings. But in the main, concepts of ergonomics can help you save money.

Fresh insights with ergonomics glasses

One of the greatest values of ergonomics is that it can cause people to think and thus promote innovation. We can put on our "ergonomic glasses" and view the workplace and end-products from a new perspective. We begin to ask questions about how a tool or production process ought to be designed to make it more human-compatible. Ultimately, that thought process can stimulate fresh insights on old problems. Ergonomics can be a tremendous source of innovation.

Through our ergonomic glasses, we can spot problems that we have overlooked before. We can challenge assumptions, find new ways to accomplish our goals, and sometimes find tasks that simply do not need to be done anymore.

Involving the user

Creativity occurs when end-users and designers interact, which is part of the ergonomics process. In the workplace, this interaction can involve team efforts between managers, engineers, and employees. For development of products and services, it can mean involving customers in ergonomic studies that aid in design. In each case, we can gain insights into products and production as we tap each source of ideas, learn from each other, and spark new thoughts.

40,000 years of ergonomic progress

Another perspective that can help us think about ergonomics in a way that makes good business sense is to realize that in many ways humans have been using ergonomics for roughly 40,000 years. In the same way that a chemist can view much of the world as chemistry, we can see human advancement as ergonomics. In this sense, ergonomics has existed ever since the first human picked up a stone to use as a tool, capitalizing on a human capability and overcoming a human weakness. Adapting our surroundings to fit us is indeed one of the defining characteristics of our species.

Great Ergonomic Improvements in History

Improvement	Ergonomic Benefit
Stone ax	Greater strength
Wheel	Reduced exertion
Chair	Improved posture; reduced static load
Power sources: water wheel, steam, electricity	Reduced exertion and repetitive tasks
Computer	Reduced repetitive tasks
Sliced bread	Reduced motions

Closer to our own daily lives, we've all had experience with ergonomics, even if we have never used the word. We tend to arrange things to fit our own convenience, at least when we can. We keep things we need within easy reach, we change our posture when we are tired of being in the same position, we shift to avoid glare. We try to modify our surroundings to make things easier for us.

So, in some ways, ergonomics is nothing new. By definition, anything that improves (or has *ever* improved) the interface between humans and systems is ergonomics. Humans have always tried to find better ways of working, taking advantage of our talents and using tools and machines to overcome our limitations. So, from this perspective, it should not be news that ergonomics can promote progress and productivity. As stated above, as a species, we've been doing this for quite some time.

Systematic process

The point is that ergonomics is not necessarily anything esoteric or extravagant. On the contrary, much of our economic and technological development has been "ergonomics."

What *is* new, however, is the scientific approach to understanding human anatomy and physiology and then methodically designing for people. What now lies before us is to take this natural tendency and turn it into a conscious approach to management and a systematic process for design.

Design failures

Altho ergonomics is in many ways a human tendency, we don't always do things the right way. A special problem is that many times the designers (and here we include anyone who has set up a task or built a piece of equipment) are not the users, so they may not know the unintended consequences of their plans.

All too often, we plan workplaces based on "efficient movement of product" or "best locations for machines," all without much thought to how people are supposed to fit in. Too often, we devise products based on the cheapest way to manufacture, or perhaps on aesthetics, but without much regard for the end user. We *think* we are paying attention to the bottom line, but we may be missing important costs, such as injuries, errors, and inefficient motions.

Although designers often think about how to fit the task to the person, clearly at other times they — that is, *we* — do not. Consequently, a formal ergonomics process is essential for business.

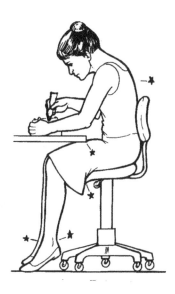

Poor ergonomics can cause discomfort and injury to employees, plus be inefficient.

Hidden costs

Too often we expect people to adapt themselves to fit into whatever arrangement has been devised, believing that it has no associated cost. Unfortunately, the human body cannot adapt to everything. People have differences and they have limitations. If business does not understand the basic requirements of humans — how far we can reach or how we perceive information — then there can be many unnecessary costs and failures.

There can be inefficiencies in production and human error in product use. There can be frustrated employees who quit and customers who switch to a different brand. There can be injuries and workers' compensation costs and product liability suits. There may be countries whose markets are closed to our products simply because we ignore differing statures of people. The price tag for design failures can be staggering.

**Analogy:
long car trip**

Perhaps an easy way to understand how something as simple as a good chair can affect productivity is to think of when you have driven a car with the seat adjusted for someone of a radically different height. Almost everyone has done this on occasion and has quickly become pained and fatigued. When you continue in this constrained posture, you ultimately reduce your attention on driving or you need to stop driving and stretch.

We would not think of buying a car without good seats and adjustability, even though few of us drive cars eight or more hours per day. If we went to a show room, we probably wouldn't consider a car without adjustable seats, even if the price were much less, and especially if the seat had no cushioning.

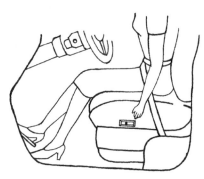

Similarly, when workplace equipment does not fit, we become pained and fatigued and we lose focus on our work or stop work altogether. It is curious that we do not think of the workplace in the same light as we do cars, even tho we spend more time at work than driving, and despite the fact that for most people working is more important financially than driving.

**Every business
can use it**

Ergonomics has broad applications that can be used in every business. The opportunities for improving the interface between humans and systems surround us. Every employer can use ergonomics to improve the fit between the place of work and the people who work there. Every producer of goods and services can use ergonomics to enhance the fit between the product and the customers who use it.

Businesses are all about people — how best to use people to make products and services that best help the customers who use them. At every juncture there are people interacting with tools and systems. And at every point in this web of interactions, we can use ergonomics as guidelines for improvement, benefiting business.

This is true whether the business is a one-person firm working out of a home or a multinational corporation with an international network of plants and sales outlets. It is true for all organizations whether for profit or not; it is true for the public sector as well as private.

The ergonomics edge

Ergonomics can enhance the functioning of any organization and can, in fact, be a formidable tool to gain a competitive advantage.

People as assets — We often hear the slogans: "people are our most important assets," or "in the end, the only source of competitive advantage is people." These statements *do* center on truths and hit the core of the competitive advantage of ergonomics. The field offers an approach to the design of work that is based on people — our differences, our limitations, and our reactions and expectations. It offers a process and a way of thinking about the workplace that can turn these slogans into a concrete system of management.

The user-friendly workplace — User-friendly means that things are easy to understand and apply, that mistakes are reduced, and that the human is treated well in the process. It should be self-evident that anything that makes the workplace more user-friendly is good for the bottom line. You can ask yourself how a workplace that is unfriendly could possibly be more efficient than one that is.

Protecting your human resources — The smaller your organization, the greater risk you have of disruption due to the loss of a key person — even if this loss is just a few days because of back pain. Furthermore, you may not be able to replace that person, even temporarily. Even if a replacement is found, skill levels may not be the same or the learning curve may be extended before a new hire is up to the speed of an experienced employee. The more skilled that person is, the more difficult replacement becomes. The more they know about your operation, the more expensive the loss.

Improved morale — Discomfort, aches, pains, and frustration caused by inadequately designed equipment and workstations can easily affect morale. Often it is the little things that frustrate employees and create dissatisfaction, for example the hard edge on a piece of equipment that the employee continually bumps into and no one will fix. These are the kinds of issues that can emerge with a focus on ergonomics and can often be resolved relatively cheaply.

Estimating costs for poor morale is a notoriously difficult endeavor. However, almost everyone in business recognizes that there can be costs. Productivity, absenteeism, and turnover can all be dramatically affected. Applying good ergonomics shows a concern for employees and their well-being that can produce a payoff in improved morale.

The core of work

Ergonomics addresses the core of work: how we humans interact with the tools and equipment we use and the tasks we perform. Whenever the relationship between a human and task is made more effective, that's ergonomics, and it's obviously a benefit to business.

The Costs of Doing Nothing

Preventing injuries is sufficient justification for taking steps to implement ergonomics improvements. But additionally, there are important financial costs that can be reduced or avoided. These costs are often hidden or difficult to obtain in specific instances when making cost-benefit decisions, yet they are factors. The primary costs of neglecting ergonomics problems are the following:

High workers' compensation costs

The high cost of workers' compensation insurance has been a primary motivator for employers to initiate programs in ergonomics. The tools and concepts of ergonomics have been particularly effective in reducing the primary sources of the most costly injuries: back problems, wrist disorders, and assorted strains and sprains.

Most employers recognize that workers' compensation costs are rising. Unfortunately, many managers still either accept these costs as a part of doing business, or assume that the only way to seek relief is through their state legislatures. Employers must learn that they have many options for controlling these costs directly, including preventing workplace injuries from occurring in the first place through good ergonomics programs.

Awareness of these rising costs can be used to good effect when justifying purchase of new equipment or renovations of work areas. Often the addition of these costs into the cost/benefit equation can tip the scales in favor of making the improvements.

While this message is aimed at employers, it is also true that employees and unions need to be concerned. As workers' comp costs take a larger share of the payroll, it means that less money for wages and benefits is available for employees.

Taking action is easier in self-insured companies, where the link between injuries and out-of-pocket costs is more direct. But even smaller companies can reduce premiums by achieving a better rating than the industry average.

Cost breakdown

Analyses of workers' comp costs that I have conducted in a number of companies (large and small) show the following:

- Musculoskeletal disorders (MSDs) typically account for about one-third of workplace reports of injury. But more importantly, they often account for about 75% of costs.
- The costs of various types of MSDs that require surgery approximate the following:
  - Wrist disorder: $15,000
  - Shoulder injury: $20,000
  - Back injury: $40,000

Altho these are rough averages, you can use them to help frame cost/benefit decisions in your workplace.

Additional costs

Workers' comp claims are the *direct* costs related to the conditions that cause workplace MSDs. But inadequate attention to the human factor in the workplace can lead to a variety of additional costs as well, often referred to as the *indirect* costs:

Turnover — Dissatisfaction caused by fatigue, working in uncomfortable postures, and experiencing symptoms of MSDs may easily lead to increased employee turnover.

Absenteeism — A common reason why workers are absent is that they are experiencing early stages of an MSD. Work that hurts doesn't exactly encourage people to come every day.

Mistakes — People working in awkward and uncomfortable postures are not in a position to do their jobs right the first time. Mistakes are more common.

Restricted duty — Finding alternative work for employees who are on medical restrictions because of MSDs can cause considerable disruption in the workplace. Special treatment for employees on restrictions can cause the resentment of other employees. In some workplaces, it can be difficult and time-consuming to even *find* work that is suitable.

Paperwork burden — The red tape involved in handling MSD cases can also entail significant staff time and costs.

The 4:1 rule

Many companies have found that multiplying their direct workers' compensation costs by four (4) provides a good estimate of the indirect costs For example, a $20,000 cost for shoulder surgery multiplied by four yields an estimated total cost of $80,000. You can justify a lot of changes and new equipment for $80,000.

Some companies that have studied their indirect costs in detail have discovered that this ratio is higher. I am familiar with a poultry processing company with high turnover where it was actually 14:1.

On occasion, these estimates start generating such large numbers as to lose credibility. I have found that just doubling the direct workers' compensation costs often yields estimates that are sufficiently large to justify considerable investment in ergonomics.

The fatigue rule

Each minute of productive time lost because of fatigue and discomfort costs roughly $100 per year (based on approximately $40,000 per year in salary and benefits). If better tools and equipment save just five (5) minutes of human down time from fatigue and discomfort per day, the amount saved is about $500 per year.

Multiply $500 by the number of the employees who are affected by a proposed improvement that reduces fatigue and it can help justify investments. As an example, this approach can be helpful in justifying better office chairs.

The profit margins rule

If a company has 10% margins of profitability, then by definition, it takes $10,000 in sales to generate $1000 in profits. Conversely, if there is a $1000 loss (due to an injury or whatever), it takes $10,000 in sales to make up for that loss.

Consequently, this approach can be helpful in getting the attention of people who you are trying convince. Use your own workers' compensation losses and profit margin. In essence, you *divide* your losses by the profit margin to obtain the sales that are necessary to recoup the losses.

Take the broad view

It is usually not possible to justify every single improvement, both because of the intangibles in costing human resources and the difficulty in estimating future losses that might be prevented. For example, it is difficult to prove that buying a specific pallet lift will prevent a specific injury.

It is much easier to justify the investment in better equipment in a setting that has a history of problems and a record of high workers' comp and other costs. Ironically, justifying expenditures is much more difficult when you are in a facility with a good safety record and trying to continue to be proactive. That contradiction points to the inadequacy of the standard approach to cost justification.

Furthermore, it is virtually impossible to calculate the financial benefit of every little item such as an anti-fatigue mat or a raised platform for a particular employee. For that matter, it can be difficult to cost justify a drinking fountain or paving the parking lot or any of a number of items that we routinely provide.

Often it helps to take a broader view and make judgments based on the sum of the arguments made in this section.

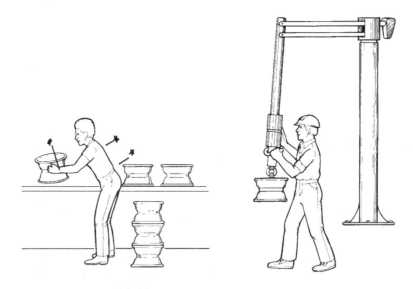

Employers should not assume that high workers' comp premiums are simply a cost of doing business. You *can* control your workers' comp costs with good ergonomics.

Studies

There is a considerable amount of data available today that demonstrates the positive financial results from good ergonomics, much of it accessible on the internet. It is not the intent of this book to provide a review of these data. However a few other studies have special value and ought to be included:

$30 billion/year potential savings

A major study performed by the Government Accounting Office (GAO) on the experiences of a number of major corporations with comprehensive ergonomics programs. The average decrease in workers' compensation was over 60%.[1]

Workers' compensation costs to employers nationally total over $50 billion per year. These costs amount to 1.3% of payroll. These numbers have decreased modestly in recent years because of (1) aggressive employer safety and ergonomics programs and (2) administrative changes in state workers' compensation laws and regulations.[2]

Combining these findings, you can conclude that if all companies set up ergonomics programs as effective as those in the GAO study, it would save U.S. employers a massive $30 billion annually.

$61 billion/year loss from pain

In addition to workers' compensation, lost productive time from common pain conditions among active workers costs an estimated $61.2 billion per year. Most of the lost productive time (76.6%) is explained by reduced performance while at work and not work absence.[3]

25% higher office productivity

Most productivity studies are in the form of before-and-after comparisons, which are usually sufficient to help management decision-making. However, they are not very rigorous from a scientific view, since there usually is no formal control group.

One of the few studies in a controlled environment was done in an office setting. The study found a roughly 25% increase in output as well as an improvement in subjective feelings of well-being. In this study, temporary employees were hired to do data entry, and were shifted back and forth between regular offices and ergonomically designed offices.[4]

---

[1] *Private Sector Ergonomics Programs Yield Positive Results*, U.S. Government Accounting Office, GAO/HEHS-97-163, 1997.

[2] *Workers' Compensation: Benefits, Coverage, and Costs, 2001*, Study Panel on National Data on Workers' Compensation, National Academy of Social Insurance (NASI), Washington, DC., July, 2003.

[3] Stewart, W.F. et al. Lost productive time and cost due to common pain conditions in the U.S. workforce, *Journal of American Medical Association*, 290: 2443-2454, 2003.

[4] Dainoff, M.J. Ergonomic improvements in VDT workstations: health and performance effects. In *Promoting Health and Productivity in the Computerized Office*, ed. Sauter, S.L., Dainoff, M.J., and Smith, M.J. Taylor & Francis, 1990.

Human Resource Trends

Several long-term workforce trends may affect the profitability of many companies. Or to make the point more forcefully, if employers do not take these factors into account, the effects might be costly to the company as time progresses. Ergonomics can help management adapt successfully to meet these changes.

Aging workforce

A larger percent of the working population is older than ever before. This trend is a result of baby boomers moving into our older years, plus better medical care and longer life expectancy. Consequently, a large share of the workforce is experiencing the limitations of advancing age. Good programs in ergonomics are needed to compensate for the increased limitations of aging employees.

Employers who have downsized in recent years have often experienced this effect most severely, since layoffs typically affect younger, low-seniority employees first and only older workers remain. Some companies have accepted that their current workforce is the one which they will have for quite a period of time, growing older as a group each year. These employers, in particular, need to plan for accommodations that allow these employees to perform efficiently and safely.

While there are many advantages to having a highly experienced workforce, the concern here is for reduced physical capabilities. Applying the principles of ergonomics can counteract these limitations. Examples include:

Poorer eyesight — Improve lighting and clarity of signs, displays, etc.

Less agility — Improve heights and reaches and provide more adjustable equipment.

Reduced strength — Reduce force requirements of tasks.

Increased susceptibility to some types of MSDs — Reduce the MSD risk factors: awkward postures, force, static load, etc.

Good programs in ergonomics can offset the limitations of aging employees. Older employees have more experience, tend to be more reliable, and are already trained and educated. When ergonomic adaptations are made, older workers can be as productive as younger workers, if not more so.

Issues that were not problems in the past might be so today because your workforce may be older.

Women in the workforce

The increase in the percentage of women in the workplace has already made its impact for the most part, but is included here to emphasize the need to modify equipment and tools for women, who are often smaller-statured.

It should always be emphasized that generalizations based on issues such as gender have not proven to be of much value. Capabilities of individuals are what matters. For example, in the case of strength we often think in sweeping (and erroneous) terms of all men being stronger than all women. We tend to forget that many women are clearly stronger than many men. Thus, for individual women, no particular modifications need be made, while accommodations may be required for some men.

Individual differences aside, women do tend to be smaller-statured than men. Ergonomists often hear complaints from women that tools and equipment are too large for them to do their jobs well. Thus ergonomic modifications are needed to enable many women to perform to their full capacity. For example:

- Tool grips should be adjustable or available in several sizes.
- Work benches may need to be adjustable or have standing platforms available.
- Long reaches may need to be reduced.
- Heavy exertion requirements of many tasks may need to be improved.

Again, these changes should not be thought of strictly for women. By making universal design modifications, all human performance becomes more effective and productive.

Rising health care costs

Health care costs in general are continuing to rise and are a factor that has clearly affected corporate profits. Preventive medicine — in this case, ergonomics — can help reduce these costs in two key ways:

- It is not unusual for people *not* to file workers' compensation claims for work-related disorders if the employer's health insurance pays for the treatment. There are several reasons for not filing a claim: the employees may not recognize that their problems are at least in part work-related; they do not wish to be perceived as troublemakers; or they may view the workers' compensation system as too complicated or time-consuming to be worth the effort. Nonetheless, they incur costs — costs that the employer pays and that can be reduced through good workplace prevention efforts.
- The health care system pays for off-the-job MSDs, which may be even more common and expensive than on-the-job MSDs. By providing training and emphasis on ergonomics at work, employers can lead people to adopt better methods and use better equipment and tools at home, thus reducing costs to the overall system.

Low unemployment

At the time of the second edition of this book, the American economy is recovering from a long downswing of the business cycle and it may seem odd to refer to low unemployment as a problem. However, at the time of the first edition, the economy was soaring and many companies had difficulty in recruiting. Furthermore, even during a national slump, certain regions can prosper, leaving at least some companies with difficulties in attracting people with the types of skills they need. Consequently, it is still important to raise this issue.

When the economy is good, low unemployment may require employers to make changes merely to attract a workforce. When jobs go begging, who wants to work in one that makes you hurt?

The best example of this effect occurred in Scandinavia in the 1970s, when unemployment dropped below 1%. Employers in those countries, especially Sweden, responded by investing heavily in ergonomics to attract and keep a workforce. This modernization enabled many employers there to continue to compete successfully in the world market (and helps explain why the Scandinavian countries lead the world in ergonomics).

In the U.S., there have been times recently when unemployment was so low in certain geographical areas that some companies had to close second shifts — a costly step — because they could not attract enough qualified workers. Once again, this dynamic shows the business value of good ergonomics. Improvements in the workplace may be needed to help attract a quality workforce.

Changing demographics

Closely related to the "problem" of low unemployment is the changing demographics of the workforce. Some employers have in the past relied on a steady stream of young males entering the workforce to do physically demanding jobs. This option may not be available in the future, if it is even now. Creative application of ergonomics can reduce the physical job requirements in a way that keeps efficiency high.

Higher expectations

Today's workforce arguably has a higher set of expectations about work than previous generations. It is probably fair to say that our grandparents and the generations before them expected work to be somewhat unpleasant and grueling. People today do not appear to accept the prospects of coming home at the end of the day worn out and hurting. The comforts and standards of the home environment have improved in recent decades and one would anticipate that most people would expect a parallel improvement in the work environment.

Again, this is hardly a startling revelation. But highlighting this shift reinforces the value of ergonomics. By applying the principles of ergonomics to all tasks, we can design the workplace to help meet rising expectations for comfort and ease.

Actions on cost justification

There are a number of specific steps you can take to help justify investments in improved equipment, whether purchased or modified in-house:

1. **Know your workers' comp costs.** Incorporate them into capital acquisition requests or work orders to help justify new equipment or modifications. In the past, most cost justifications were based on only a few variables such as reducing labor or improving efficiency. Adding a factor for workers' comp provides more realism to the equation and often makes it possible to justify equipment that may have been on a wish list for some time.

2. **Estimate indirect costs** or calculate them when possible. Once these hidden costs are known, it is often much easier to justify changes. On occasion, the magnitude of these hidden costs can be stunning and improvements that ordinarily would be considered quite expensive are easily justified.

3. **Charge costs back to departments**. Rather than have department management think of workers' comp costs as corporate overhead, these costs should be charged back to the department budgets where the injuries occurred. This practice can provide a truer picture of costs and can markedly change how decision-makers think about prevention.

4. **Remember that *qualitative* statements** are sometimes enough to obtain support. Most good managers recognize that not all costs can be quantified, especially for safety and human resources. Mere recognition that these costs do occur can be helpful in making changes. Even without being able to obtain hard cost data, reflecting on these issues can often show that ergonomic improvements can save money.

5. **Add the term "safety"** to the capital acquisition request. Many decision makers are reluctant to turn down legitimate requests that involve safety consideration.

A better understanding of the true costs of your current methods can help justify the investments for change.

A People Program

Ergonomics can help both public and private employers
in a couple of especially people-oriented ways:

Empowerment and involvement

Ergonomics fits well into current efforts to involve and empower people at work. The process of applying ergonomics in the workplace takes advantage of employee capabilities, ideas and input. Many ergonomics issues can, in fact, only be addressed through the active participation of the employees who do the actual work.

Empowerment means not only allowing people to make decisions, but also providing sufficient training to increase their competency in making those decisions. For example, millions of employees and supervisors are responsible for laying out work areas and establishing work methods, and yet few have been trained in the principles and techniques of doing so correctly. Teaching people about ergonomics can help fill that gap. Moreover, being able to apply concepts of ergonomics is a skill that employees can bring to any task, and thus a valuable asset in our rapidly changing technological environment.

Programs already established in the workplace that involve people can be used as a base for ergonomics. Examples include employee involvement in quality or improvement. Ergonomics can take advantage of these previous efforts and, in turn, contribute back to them.

Conversely, if a company has never established these formal mechanisms to involve employees, focusing on ergonomics issues is a good place to start. The concepts are relatively simple and result in direct benefit to the employees themselves, which both serves as positive reinforcement for contributing ideas and provides a base for expanding to other issues.

Comfort

Sometimes people have gotten the simplistic impression that ergonomics means providing "cushy" jobs or slowing down production. "We pay people to work" goes the saying, "not to take it easy."

Reducing *discomfort* may sound acceptable, since it connotes removing barriers to human performance. But trying to provide *comfort* doesn't sound right in the harsh world of business. However, a closer look at the word sheds some interesting light.

The English word *comfort* is taken from Latin and literally means "to strengthen" ("*com*" is an intensifying prefix; "*fortis*" means "strong" or "force"). Other modern words taken from this Latin root include: "fortify," "fortitude," and "fortress."

For whatever reasons, the meaning of *comfort* changed thru the centuries from meaning "to strengthen" to a softer meaning of "to console" or "to ease." But it is the original sense of this word that captures better what is at stake: "to *strengthen* the ability of a person to perform a task better."

Union relations

Ergonomics issues are often good ones for joint problem-solving between management and labor. Redesigning the workplace using the principles of ergonomics is a "win-win" situation for management and labor. From the union's viewpoint, jobs are improved, injuries are decreased, people are involved and become more satisfied. These are also worthy goals from management's viewpoint, in addition to reduced costs and increased efficiencies and innovations.

Experience in many industries shows that after starting joint union-management programs on basic issues such as worker safety and workstation design, new relationships were established with positive impact on other areas. Joint programs in ergonomics can thus pave the way to other joint problem-solving efforts.

Labor relations turnaround

In my own experience in the auto industry in the 1970s, I witnessed a turnaround in relationships in many locations — from bitter adversarial ones that can only be described as "lose-lose" to relationships that fostered effective joint problem-solving and even mutual respect. The initial effort to improve these relationships was concern for worker health and safety, which began to expand into workplace ergonomics in the early 1980s. Far-sighted leaders in both the UAW and the Big Three had agreed to find ways to change old patterns. Workplace safety was the common ground from which other efforts grew.

This is not to suggest that unions and management always have the same interests; they clearly do not. However, on many issues they do share a common agenda, and working on these issues together can help to develop improved relationships on all issues. Both parties, however, must see their common interests are being served if they hope to improve a working relationship.

Industrial relations consists of a web of interactions that shift and change, sometimes daily. On some issues, interests intertwine, and on other issues or at other times, interests may conflict. Finding these moments when interests are in harmony and having ergonomics tools available to take advantage of the opportunities is a key to success.

Ergonomics in Products and Services

Most of this book is written from the perspective of the workplace. But much applies from your customer's point of view for the sale of your products and services.

Customer appeal

It should be self-evident that increasing the friendliness of any product or service — improving comfort, reducing exertion requirements, eliminating confusing controls, and so forth — improves the customer appeal. A better design achieved through good ergonomics can provide a tremendous edge over the competition.

There are almost always ergonomics issues involved in the use of products, which when ignored, are sometimes fatal to the product. Note that these concerns do not necessarily involve musculoskeletal injuries, rather center around a broad set of questions that have to do with human-product interface:

1. How is a human involved with this product?
   - Is it easy to use?
   - Is it easy to service and maintain?
   - Is it easy to install?
   - Is it easy to learn how to use?
   - Is the way that people actually use the product the same as you intended?

2. Are there physical issues?
   - Exertion, reaches, contact stress, lack of clearance, etc.?
   - Vibration or noise?

3. Are there cognitive issues?
   - Confusing instructions?
   - Non-intuitive operation of controls?
   - Knobs and dials not standardized?
   - Long learning time?
   - Poor labeling or signage?

Formal review

Every product should undergo some type of review while "wearing your ergonomics glasses." This evaluation can be simple or very complex depending on the needs, but it should be done formally.

Chapter 2
How Ergonomics Promotes Efficiency

The tools and insights of ergonomics can help eliminate wasteful activities in multiple ways. Indeed, everything in the previous chapter has to do with efficiency and eliminating waste. Musculoskeletal injuries, workers' compensation costs, and absenteeism are all wastes. Their elimination or reduction yields increased efficiency.

But there is more to be said. To help make the case for why organizations need to apply the workplace ergonomics process, we need to emphasize its benefits for efficiency.

Fundamental points

There are several essential points regarding the relationship between ergonomics and efficiency. Each of these is described in more detail in this chapter.

1. The principles of ergonomics apply as much to promoting efficiency as they do to preventing injury. In fact, we might see the "risk factors for MSDs" simultaneously as "risk factors for inefficiency."

2. The activities that are hardest on people tend to be non-value-added actions — bending, reaching, lifting, pushing, pulling, carrying, orienting, etc.

3. A good ergonomics task analysis penetrates into operations step by step, very similar to old-fashioned time and motion analysis. By evaluating items such as fatigue, motions, and exertion through a task step by step, it is possible to identify wasted activities.

4. Ergonomics is an essential ingredient for other methods of improving workplace operations, such as lean manufacturing or design-for-assembly. You cannot apply these other processes well without ergonomics.

5. When concepts of *productivity* and *efficiency* are properly understood, there need be no conflict with employee well-being.

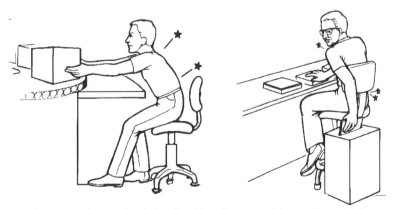

Long reaches and twisting (and bending, repetitive motions, high force, etc.) can be inefficient and wasteful, as well as injurious.

The Human-System Interface

To the founders of the field of ergonomics back in the 1940s and 1950s, it would undoubtedly seem odd to need to argue that ergonomics can be a tool for improving efficiency. That was the whole point of their efforts.

Military aircraft

The term *ergonomics* was coined by aircraft designers for the British Royal Air Force in the Second World War. Their goal was to create a cockpit that was more human compatible, so that the pilots could physically reach all the knobs and switches as well as understand the increasingly complex array of dials and indicators.

Their objectives had nothing to do with preventing back injuries or carpal tunnel syndrome. It was all about efficiency, using this special focus on studying the interface between humans and systems (or, to use the jargon of the time, "man-machine" systems).

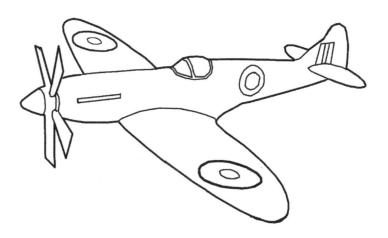

In the early days of ergonomics, the focus was on efficiency. Only recently have people realized that the same concepts could reduce the risk for MSDs.

Maximizing human performance

More recently, people have discovered that the principles of ergonomics provide exceptional value in the workplace for preventing MSDs. But historically speaking, this is a side effect.

In my view, reducing injuries in and of itself is sufficient reason for many (if not all) employers to adopt ergonomics into their everyday worklife. But the field provides so much more value that "merely" injury prevention.

The point of ergonomics is to maximize human performance, whether by reducing injuries or by increasing capabilities thru better design. It is crucial to underscore this larger purpose.

Risk Factors for Inefficiency

People involved in prevention of MSDs have in recent years learned to think of repetitive motions, high force, awkward postures, etc., as *risk factors* for injury. These factors might not be causing injuries in every case, but they certainly raise the risk of doing so. Equally important is seeing these same factors as potential sources of inefficiency.

In fact, borrowing the medical terminology of "risk factor" might not be a bad way of thinking of these issues. For example, when you spot an employee working in an awkward posture, something might be wrong and there may be a possibility or "risk" that the set up is not efficient.

As we will see in later chapters, part of the task evaluation process is to study the various issues like bending and reaching to see if they are causing physical problems for employees, then brainstorm ways to make improvements. These issues — the bending and reaching, the repetitive motions, the static gripping, etc. — serve as red flags for further investigation. The red flags could indicate a possible contributor either to a wear-and-tear type injury or to inefficiency . . . or more likely, to both.

Awkward posture

Working in awkward postures can directly reduce efficiency in three ways:
1. **Reduced strength** — Think of bending at the waist and reaching out across a large object and then trying to exert. You have little or no strength in an outstretched position like this. Consequently it takes you longer to complete a task than it would be if you were working in a proper position.
2. **Less accuracy** in your motions — Again, think of reaching out across a large object and trying to do something intricate. You make a lot of mistakes and it takes a lot longer time, if indeed you can do it at all.
3. **Faster fatigue** — When you work in an awkward posture, you tire much more easily, which slows you down.

High exertion

In general, the more exertion it takes to perform a task, the longer it takes. Think of opening a jar of jam. After you've opened the jar the first time and the lid is still clean, you can remove the lid with an easy twist. But once the jam starts to build up on the lid, it requires more force and it takes longer time. Likewise, a heavier box takes longer to move than a light one and a screw that is hard to turn takes longer to insert than one that moves easily. A drawer that sticks might take you two or three attempts before you can open or close it.

Moreover, muscles under a heavy load are harder to move with precision. Thus accuracy of movement is reduced, which has consequences for both quality and efficiency.

Fatigue

There is a direct link between fatigue and lower productivity. Fatigued people produce less and make more mistakes.

The efficiency experts in the early days of mass production understood this negative impact and devoted time to studying causes and effects. However, issues like fatigue have become neglected in our time, possibly because of the lure of the more exciting high-tech developments. But the issue still remains important and is gaining visibility again because of ergonomics.

Fatigue is one of the key areas of ergonomics study. In particular in recent years, the issue of *static load* on muscles has become the predominant focal point of workplace activities, but heavy, exhausting work is certainly still a problem in some industries.

Wasted motions

Repetitive motions are a waste of time . . . literally. The more motions, the longer it takes to perform a task.

Nearly 100 years ago, the pioneers of scientific management developed a technique called Time and Motion Study. By the end of the 20th century, that technique had mostly degenerated into just Time Study.

Now motion analysis is coming to the fore again. A good ergonomics analysis seeks to identify the types of motions required for different steps of the job. With such focus, it is possible to identify instances where it is possible to improve the type of motion being used or reduce the number of motions, if not eliminate them entirely.

Repetitive motions should be viewed not only as a source of injuries, but also as a red flag for wasteful work.

Heights, reaches, and more

Poor heights and reaches can affect productivity in a couple of different ways:
1. If you can't reach an object at all, you may need to stop productive work and fetch a step stool or take time to remove an obstruction.
2. If the inappropriate height or the long reach causes you to work in an awkward posture, you end up losing productivity for that reason.

Contact stress can create discomfort, which in turn can reduce productive work. Lack of clearance can cause awkward postures. Vibration and noise are contributors to fatigue. Excessive heat or cold can clearly inhibit human performance.

The list goes on.

Point

The point that bears repeating is that when you look at a job with your ergonomics glasses, you should think about how the issues contribute to inefficiency as well as injury.

Non-Value-Added Risk Factors

The activities that are hardest on people tend to be non-value-added actions — bending, reaching, lifting, pushing, pulling, carrying, orienting, etc.

Adding value

The term *value added* stems from the concept of *the market value that is added* to a product or material at each stage of its manufacture or distribution. For example, think about attaching a metal buckle to a leather belt. The buckle and the leather strap are each worth so much, but when you attach one to the other, the combination is worth more to the consumer. You have added value to the product.

Non-value-added steps are everything else: conveying the belts and the buckles to the assembler, putting the completed belts in a package, and taking the packages away. When the assembler takes the step of reaching into a container to pick up a buckle, that step does not increase the worth of the product. It might be necessary, but it doesn't add anything from the point of view of the customer.

We would want to eliminate waste in either case, whether value-added or not. The distinction is made to help focus on what is crucial and what isn't

MSDs

The point here is that it is common in industry for the non-value-added steps to involve much of the physical work that employees perform, and consequently, have higher risk factors for MSDs. To be sure, the value-added step sometimes involves factors that increase risk for injury, but mostly the problems are in the non-value-added steps.

Example

The following illustrates the point. The method used here is and elsewhere is this book is called Time and Physical Demands Analysis. In this case, the job was running a mechanical press. A videotape of the job was paused each 0.5 second and an estimate made of strain on the employee. The results show how much physical demand there is on the employee for each step of the job.

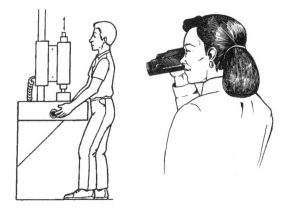

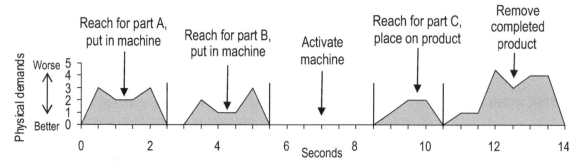

Understanding the graph

The graph shows that the cycle of making one product at this workstation takes 14 seconds. It takes 2.5 seconds to handle part A, with a slight peak when the employee reaches farthest out to pull the part from a container and another peak to place the part in the machine (often, we would consider these as two different steps, but for simplicity's sake, it is treated here as one).

There is a half a second of moving, which involves no strain on the body in this case. Then there is another 2.5 seconds to handle Part B, which happens to be lighter than part A, but involving the same motions. Consequently, we see another double peak, but the peaks are lower.

Next comes the three seconds that it takes for the machine to function. The machine itself is activated with a photoelectric switch, so there is no strain on the employee to trigger it.

It takes two more seconds to add another small part to the product. Then it takes 3.5 seconds to reach in the machine and remove the final product to place on a takeaway conveyor. The final product now contains the combined weights of all three parts so the graph rises highest here.

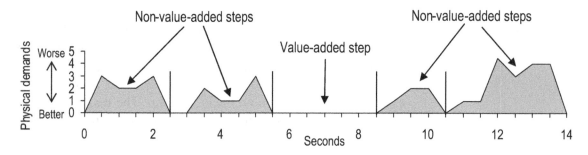

Points

The first and most important point is that all the strain on the employee is related to the non-value-added steps.

A second point is that the most strain and the most time are associated with removing the completed product. As it happens, the discussion at this workstation revolved around the feasibility of adding an automatic ejector to the machine, which would thus reduce the cycle time by about 20% and the physical demands on the employee by about 40%.

A third point is that if the containers for the three parts were moved closer, both the peaks of strain and the time involved would also be reduced.

Another example

An additional example helps to reinforce the lesson. In the following situation, the job involved carrying a heavy product from a container to a workbench, then to another workbench, and then back to a different container.

In this particular case, the time needed to perform one cycle was about 80 seconds, as shown in the graph below. Basically, there are two operations for this job. Each of these contains multiple steps, but is simplified here as Operation 1 and Operation 2. For this job, many of the value-added steps did involve activities that increased the risk for MSDs. However, the graph clearly shows that most of the physical strain came from the non-value-added steps of carrying the heavy product from place to place.

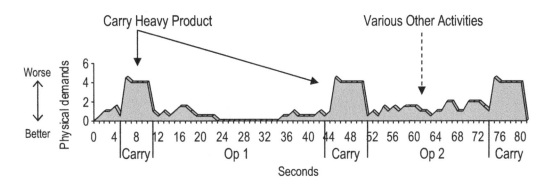

Implications

This analysis leads to the following points:

- About 16% of the work cycle was spent carrying the heavy product from workstation to workstation, that is, wasted non-value-added work.
- About 70% of the physical demands (increased risk for musculoskeletal injury) were related to carrying the product.

In this particular situation, it was feasible to rearrange the layout in a U-configuration to permit sliding the heavy product and thus eliminating the carrying. Exertion and time is still needed to slide the heavy product, but there are savings for both issues.

Common situation

As with the previous example, the problem was identified as part of an ergonomics evaluation to reduce the risks for MSDs. To be sure, any other type of close evaluation, such as time study or process improvement, could have uncovered this wasteful activity. However, it was in fact the ergonomics analysis that led to these insights. Furthermore, ergonomics analysis is particularly helpful in revealing these types of situations.

As a final comment, no claim is made here that non-value-added steps are *always* the worst portions of jobs. However, in my experience, based on conducting thousands of task evaluations, it is common and it warrants emphasis.

Finding wasted work

A good ergonomics task analysis penetrates into operations step by step, very similar to old-fashioned time and motion analysis. By deliberately focusing on items such as fatigue, high exertion, and excessive motions, the process easily lends itself to uncovering wasted activities.

Seek out demanding tasks — Every example contained in this book involving an ergonomics evaluation that led to an increase in productivity was initiated by the goal of reducing injuries. This circumstance happens so routinely that it deserves to be turned into a maxim:

To discover wasted work, deliberately seek out tasks with high physical demands.

A good example of finding waste occurred once when I was training an ergonomics team to conduct basic task evaluations. In this case, the group was conducting a walkthru evaluation and noticed an employee bending over into a bin, similar to that shown in the illustration above.

After some observation and discussion with the employee, the team discovered that the employee's machine invariably knocked the products out of alignment when they were dropped into the shipping container. Consequently, the employee had to stop productive work every so often, walk to the opposite side of the machine, and stay in this back-breaking position for several minutes to sort and realign the finished parts into their proper orientation. It was all a waste of time because of a flaw in the machine design.

Roughly 10% of the work cycle was lost to this action. Of particular note, this operation had been relocated from one area in the facility to another — and undergone an evaluation in the process — but none of the plant engineers noticed the problem.

Habits — Anecdotes don't prove anything, but they do help readers get in the habit of seeing things in certain ways. The usual problem is that anyone who stays in the same workplace for a period of time becomes so used to seeing the work being performed in the same way that obvious problems becomes "invisible."

You can make the problems visible again by using the process described in this book. Also, there can be benefit from bringing in an outsider with a "fresh set of eyes."

How Ergonomics Can Help Create a "Lean" Organization

You cannot implement "lean" well without ergonomics.

To be sure, there are some aspects of lean operations that have nothing to do with ergonomics, such as issues with transportation and vendor relations. But when an individual employee is involved (as is often the case for most lean projects on the workplace floor), ergonomics is an essential ingredient for optimal success.

The reverse is equally true: If an ergonomics perspective is not applied, the results of the lean process can sometimes lead to an increase in MSDs. The company can lose in injury costs any gains made via the lean philosophy.

Lean (or flow)

The *lean* philosophy centers on improving the *flow* of materials from vendors to manufacturers to distributors to customers. A common synonym is *just-in-time*. Within a production plant, the term *flow manufacturing* is probably more descriptive than *lean manufacturing*.

The idea is to smooth the flow by keeping parts and products moving and avoiding any need to store parts for excessive periods of time. A good flow enables companies to respond quickly to customer needs.

Sometimes the term *lean* can sound like trying to get by with the least number of people possible, with unintended consequences for employee well-being. However, the concepts are much more sophisticated than that. Inherent in the lean philosophy is that productivity is achieved by eliminating wasted activities and working smarter, rather than by simplistic notions of making employees work harder or faster.

Indeed, there are many overlaps in the ergonomics and lean processes. Both methods seek to eliminate wasted activities such as errors, defects, excessive motions, unnecessary steps, and senseless operations that are often performed "because that's the way we've always done it."

Detailed task analysis

Hopefully, this entire book helps explain how ergonomics can help promote the lean philosophy. In particular, the prominence given to studying in detail the actual tasks that employees perform can provide important insight to smooth the flow of products and materials.

One of the tenets of the lean philosophy is to have actual data on work content and to base decisions on facts rather than opinion. There is probably no better method than ergonomics for this purpose, since the whole point of the field is to study tasks in order to learn about various physical demands (like exertion, motions, long reaches, static postures, etc.). Such study includes the time and effect of each step of the job, which is important for the lean process.

Value-added perspective

It should be noted that part of the value of ergonomics is simply adding a different perspective. By coming at the job from a different direction it often becomes possible to gain new insights on old problems. You can't see everything looking at a blueprint or a diagram of a department. It helps to look at the jobs from the employee's perspective, step by step, wearing your ergonomics glasses.

Quick changeovers

Flow manufacturing often involves increasing the number of changeovers of machines and equipment. Rather than producing large numbers of products or parts and using massive amounts of time and expense to store them, the lean concept is to produce only as many items as needed, then switch to a different product.

One consequence is that more frequent changeovers entails that a task previously done perhaps only once a week may now be performed several times per day. If that changeover involves awkward or hard-to-do steps, it may not have mattered much previously because it was done infrequently. But if it is now performed more often, the changeover task may create problems. Thus it may require an ergonomics evaluation and improvements to avoid unintended repercussions.

Simultaneously, reviewing the changeover from an ergonomics perspective can help identify smarter methods for that changeover. Examples of improved ergonomics that have cut time requirements for changeovers include:

- Using quick-release latches for panel doors (reduce motions)
- Rerouting pipe lines to give better access to valves (provide clearance)
- Using permanently mounted supports to lift and remove routinely changed heavy items (reduce force)
- Using portable booms to lift heavy items (reduce force)
- Designing equipment so that needed items or heavy items are not buried deep within the equipment (provide clearance)

Reducing maintenance downtime

Similarly, ergonomics evaluations of maintenance tasks can reduce the length of downtime. There is little difference in performing a task analysis of a maintenance employee and a production employee. You still break the task down into steps and identify the same "risk" factors. You still need to think and challenge the assumptions of "the way we've always done it." The only difference is your rationale for making changes.

The ergonomics of maintenance tasks is often referred to as *maintainability*. As with other types of work, the most opportunities for improvement are when designing the equipment. However, it is still possible to retrofit equipment after it has been installed.

Typically the biggest issue is lack of clearance. Additionally, it is common for maintenance staff to be using suboptimal tools as well as needing to manhandle heavy loads.

Saving space

Many ergonomics improvements have a side benefit of saving wasted floor space. In particular, to reduce long reaches it is often necessary to provide smaller workbenches, narrower conveyors, and better layout of work cells and workstations. All of these can free up congested areas.

Cognitive ergonomics

Cognitive issues have not yet been emphasized much in workplace ergonomics programs. Activities that involve this branch of the field have focused on complex equipment such as aircraft cockpits and control panels in nuclear power plants. There are obvious safety issues involved with poor cognitive design of equipment (lack of standardized controls, unclear directions, etc.), but the impact for lean manufacturing may be even greater.

Making things understandable and user-friendly helps avoid mistakes, increase reaction times, and lower learning curves. Cognitive ergonomics is an untapped area for improving the design of machine control panels, directions, signage, forms, operation controls, sets of instructions, and so forth. All of these concepts can help create an effective and lean operation.

Overlaps in the improvement process

The process of implementing ergonomics improvements on the plant floor is very similar to that for lean operations. The manuals on how to set up the lean process read much the same as Part II of this book. Consequently, it is easy to combine the two efforts. In particular, if physical changes are to be made in production lines and work cells, this is obviously the time to incorporate all needed changes. The two activities done in tandem undoubtedly would lead to better understandings of the work process and yield better results.

Employee involvement

Employee involvement is as important for lean operations as for injury reductions. Much information about specific tasks and possible alternatives can only be known by those who perform them.

Any efforts to analyze a specific work area from a lean perspective would undoubtedly involve the same employees as for an ergonomics improvement project. Likewise, any skills gained by employees in team-based problem-solving can be applied to other projects, whether the goals are improved manufacturing flow or injury prevention.

Tapping into self-interest

An additional point to emphasize is that ergonomics provides a good way to respond to direct interests of employees. Particularly, if there have never been any formal efforts along these lines, it can be useful to start with something that is in their own self-interest (physically easier tasks), which directly rewards involvement and can set the stage for involvement in lean efforts.

Creating a culture for change

Ergonomics is an excellent way to get people to ask the questions, "Why do we do it this way?" and "Is there a better way?" A good ergonomics process starts with people asking "Why?" for each step of a job.

Creating change in an organization can often be a stubborn challenge and it helps to have a push from another direction. In this context, promoting change from the perspective of safety can reinforce efforts from the perspective of customer response time, thus building new organizational habits.

Scrap, bottlenecks, and takt time

Whenever issues of scrap and bottlenecks are addressed, it is important to determine how much of the problems are machine-related vs. human-related. If humans are involved, then there can be value in "putting on your ergonomics glasses" to understand the factors involved.

Similarly, once the takt time has been identified, then it is important to identify the human tasks that are critical and study those. There may be issues such as improper heights and reaches or wasted motions that keep an operation from meeting takt time repeatability.

Ergo Kaizen Blitzes

Some plants have taken to making ergonomics improvements as part of kaizen events, and even to perform special ergo kaizen events to get quick and noticeable resolution to problems. It makes no sense otherwise to invest the time and effort in kaizen events and not deal with those issues that may be causing musculoskeletal injuries. Furthermore, as has been argued here many times, doing the ergo evaluation will help team members see wasteful, unnecessary activities that may be associated with the task.

In many companies, you can use the same teams for ergonomics and the lean process. There are many parallels and you can multiply the strengths of otherwise separate activities.

How *lean* can support good ergonomics

The lean philosophy can help your ergonomics efforts as well as vice versa. There are a number of lean concepts that inherently improve the fit between employees and their tools and equipment. Indeed, I have been involved in situations where an understanding of lean production was a prerequisite for identifying a way to prevent MSDs.

Because of the current emphasis on lean manufacturing, it can be helpful to show some of the inherent or potential overlaps with ergonomics.

Single piece flow — A common technique for smoothing the flow of parts is to switch from the traditional method of working on parts in batches to a more assembly line style of single piece flow. Working in batches typically involves putting parts in a container, then moving the container to the next workstation to remove the parts all over again. Single piece flow essentially involves handing off a part from one employee to the next, skipping the step of using the container.

Single piece flow thus eliminates many repetitive motions in loading and unloading the containers, at the same time ending the need to manhandle the often heavy containers. Both wasted time and risk factors for MSDs are reduced.

Department vs. work cell — The traditional approach to organizing a factory has been by individualized departments (all the presses in one department, all the welders in another, etc.). This arrangement requires considerable material handling, including manual tasks.

Lean manufacturing commonly involves changing the organization so that the operations are performed in sequence (commonly called a work cell — the press directly feeding the welder, which directly feeds the next operation, etc.). This approach eliminates much of the material handling, and as with single piece flow, tends to eliminate repetitive motions.

Improving layout — Another common technique to improve flow on the production floor is streamlining layouts. Usually the method involves common sense to arrange the equipment to follow the sequential steps of the operation. Better layouts tend to make life much easier for employees by reducing reaches and removing unnecessary motions — both typical issues to be addressed from the ergonomics perspective.

Reduced Inventory — Inventory represents waste in two ways: (1) financial capital is soaked up to pay for parts that are idle and (2) the response to customers is slowed because the stored inventory must be used up before new materials are introduced into the system. Consequently, reducing work in progress (WIP) inventory is often a goal of lean manufacturing.

In many circumstances, reducing inventory can also have a big impact on reducing wear and tear type injuries. For example, having fewer parts in process tends to reduce long reaches (since having fewer parts enables smaller footprints for

How *lean* supports ergo (continued)

staging and conveying and thus less distance to reach across). Additionally, having fewer parts sometimes also reduces the need to lift heavy loads.

Even if batch production is still maintained, reduced WIP inventory often involves handling the parts in smaller lot sizes, which can entail using smaller containers that (1) are easier to move and lift, (2) can be conveyed closer to the point of operation, and (3) can reduce the long reaches and bending that are commonly involved in reaching down into overly large containers.

"Pull system" — The flow of parts in the traditional system of manufacturing is a "push" system, so called because people and machines keep producing parts and pushing them down the line whether they are needed or not. Under the new philosophy, when a machine breaks down or production backs up, people upstream stop working (anathema to a traditional manager). The idea is to not produce any parts until they are absolutely needed. Thus, parts are "pulled" through the process on demand.

The push system sometimes creates unnecessary motions, long reaches, and extra effort when the parts get piled and tangled up on one another. The pull system prevents this from happening, and if designed correctly, places the incoming part in easy reach of the employee and in the correct orientation.

Eliminating wasted motions — Many approaches to lean manufacturing emphasize the importance of studying jobs to detect unnecessary steps and unnecessary motions. De facto, such activities involve ergonomics, even if the practitioners are not consciously aware that they are doing so.

Job rotation — Lean philosophy advocates often promote cross-training and job rotation. This practice provides for flexibility in production. Furthermore, it acquaints employees with many additional tasks, which increases their areas of expertise in a way that can be extremely helpful in problem-solving and identifying improvements. Additionally, job rotation can sometimes provide benefits by reducing the risk of musculoskeletal injury.

Single piece flow can eliminate long reaches, repetitive motions, and the need to manhandle heavy containers.

Improving product quality

People who work in awkward and uncomfortable postures and who are fatigued are in no position to do their jobs "right the first time." While most of the focus on quality improvement is on machine and equipment operation, there are times when the human element comes into play.

The human factor — To err is human and can occur in a variety of ways. For example:

- Sometimes people misread information from displays or activate the wrong controls, thus causing mistakes and defects in products. The field of cognitive ergonomics is devoted to understanding ways to minimize these types of errors with better design. Cognitive ergonomics is not emphasized in this book nor has it been given much prominence in current workplace ergonomics projects. But the field is ripe for application.

- Humans are not very adept at performing the same movements accurately, plus we become distracted easily. This factor is one reason why machines and automation are useful for repetitive tasks — they are simply more accurate.

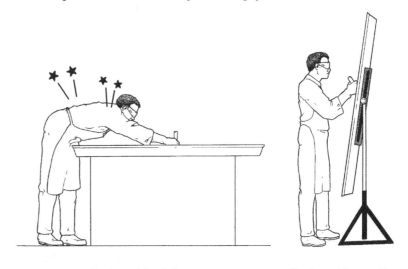

Designed for defects   Designed for quality

Physical issues — The focus of this book is on physical issues that affect human well-being and capabilities. There are routine instances where employees inadvertently create defects because they become fatigued or plainly cannot reach or otherwise perform the task. Task evaluations for these physical issues can provide the benefit of leading to increased accuracy and greater alertness.

Defective parts can trigger MSDs — The converse can also be true. For example, in assembly operations, when parts don't fit properly, the resulting increase in exertion has led to an increase in the number of MSDs.

Design-for-assembly

Another technique of manufacturing that can't be done well without knowing ergonomics is design-for-assembly. If a human is involved in doing the work, then designers must know about human dimensions, strengths, weaknesses, and the host of other considerations that make up the field.

The design-for-assembly jargon is often very different from that of ergonomics, but the issues are the same. Ergonomics provides value by putting you in the position of the assembler to spot potential problems like:

- Easy motions vs. difficult ones
- Clearance, including visual access
- Postures required to do the work
- Strength requirements
- Excessive reaches

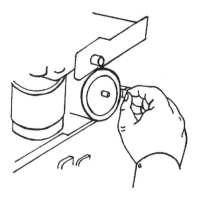

Example: If designers don't provide finger clearance, you certainly can't assemble the product easily.

The overlap between design-for-assembly and ergonomics is huge. A few examples of some of the standard rules of thumb that help show this overlap are:

- Use the fewest number of fasteners as possible.
- Use as few parts as is possible.
- Present the product to the employee in an optimal position.
- Orient incoming parts in the position that they will be used.
- Avoid difficult-to-handle materials like springs or anything else that can become entangled or can stick together.

The problem-solving process can sometimes be different, if the product is new and there no existing operation to study (that is, nothing to videotape and no employees to involve). But if the design is an adaptation or replacement of an existing product, the process can be the same as that described in this book.

Solutions

Design-for-assembly can also be a solution for a job that is causing problems for employees. Sometimes it is not feasible to improve tools or work methods and the only option is to change the design of the product to be easy to assemble.

Ergonomics in Its Place

Ergonomics provides business with an additional tool for problem-solving. The conceptual framework is improving the interface between humans and systems and it integrates seamlessly with many current initiatives in general industry.

Modern tools

Any of the modern techniques of workplace improvement, if applied intelligently, can provide benefits for product quality, efficient flow, and healthy employees. Some tools are better for some goals than others, but all have an impact.

Each of the tools lends a certain viewpoint and all combine to improve the production and distribution processes. The schematic below helps show the point:

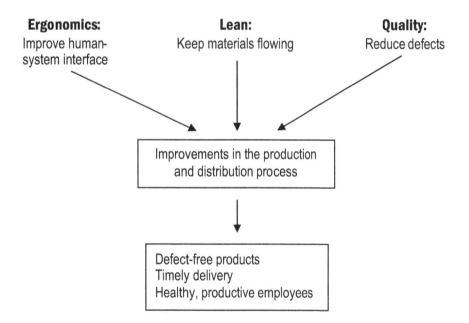

Ergonomics:
Improve human-system interface

Lean:
Keep materials flowing

Quality:
Reduce defects

Improvements in the production and distribution process

Defect-free products
Timely delivery
Healthy, productive employees

Redeemed from a bad past

This line of reasoning is made especially clear if you contrast today's workplace philosophy with that of the past. The out-dated mentality was that productivity results from getting employees to work harder and faster. This obsolete attitude was part and parcel with the view of "get the product out the door" as fast as possible without regard for quality.

The outdated system was bad for quality, bad for response time to customers, and bad for employee well-being.

Today we have seen the rise of quality, the lean philosophy, and ergonomics — all in reaction to the problems of the past. Each provides a critique of the old system, thus showing an inherent link in philosophy connecting these three newer methods. The important point is that when all the modern tools for industry are applied in concert, the results undoubtedly are better.

Perspectives on Productivity

The terms *efficiency* and *productivity* are fraught with misunderstandings and negative connotations, *productivity* in particular. Owners and managers sometimes miss the boat by focusing on the wrong factors. Employees, unions, and even the general public do likewise, but from the other side of the coin.

Leaps in productivity

Most of us realize on one level or another that increasing productivity has been the basis for the rising standard of living since the birth of civilization. Abstractly, we tend to think of the term in a positive light.

It is important to reflect on the fact that thruout the course of history there has never been a leap in productivity or a meaningful rise in standard of living that has come about because of working harder or faster. All the advancements in progress have resulted from finding smarter ways to work, taking advantage of human capabilities and overcoming human limitations.

Unfortunately, on a day-to-day basis, we often behave as though we believe that productivity simply means working harder and faster. This misconception causes problems for both managers and critics of industry alike.

Speed-up mistakes (I)

Obviously, any employer who wishes to gain competitive advantage should focus on tapping into employee intelligence rather than pushing anyone to the limits of their physical well-being. Sadly, there are still too many managers whose idea of increasing productivity is to walk around the workplace and try to identify which people to eliminate, often based solely on gut feel. But if you reduce headcount while keeping the same tasks, it can cause problems, such as increased defects in products, more rework, and more MSDs. The end result might be an overall increase in costs.

Speed-up mistakes (II)

Similarly, critics of industry have argued that since productivity has risen so much in recent decades, it must follow that individual employees must surely be more overworked than ever before. However, this inference ignores the effect of improved technology, such as increased mechanization, reduced scrap, better decision-making, better communications, and a host of other enhancements.

Critics point to speed of production as the cause of MSDs and argue that to reduce the injuries, work should be done more slowly. While it is true that faster work may have led to present problems, it does not follow that simply slowing everything down is the solution.

Just like the manager above who is dollar-foolish and penny-wise, the critics focus on the wrong things.

How ergonomics can help

The following is an example from a company that inspects and repairs bags and other small containers for a major distribution company. The union had complained that the work was too fast and should be slowed down. Management claimed that slowing down the work was not economically feasible.

Almost immediately as I started to look at the jobs, I noticed a considerable amount of bending and reaching. After a few minutes further observing the steps of the task, I realized that the demanding activity was related to an unnecessary double handling of the product and suggested a way to eliminate it. The two feuding parties asked for data to show the extent of the problem and potential benefit from changing methods.[*]

The analysis I used was the time and physical demands method that has been shown elsewhere in this book. The jobs were videotaped, and then reviewed by pausing the video every half second and calculating the loads on the lower back and shoulder using standard biomechanical formulas. The results were then graphed, which thus shows the physical demands associated with each step of the job. The following are the results for one of the jobs.

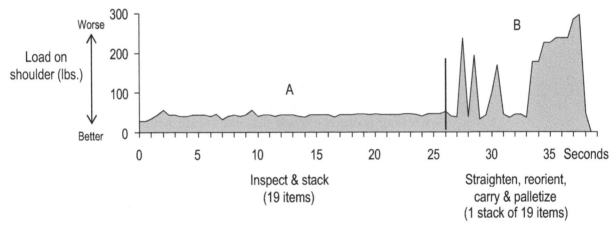

Inspect & stack
(19 items)

Straighten, reorient,
carry & palletize
(1 stack of 19 items)

This task basically involved two parts: inspecting and stacking the containers (A in the graph above), then straightening and reorienting the stack and carrying it to a pallet on the floor (B above). The argument had been about the time it took to perform step A. However, what actually caused the exertion and fatigue in the job was part B.

Moreover, it was possible to eliminate all of part B by purchasing a pallet lift and stacking the containers directly onto the pallet. This made about one-third more time available, both for more rest and for more production. Thus more containers could be inspected with less effort — more output with less input — which is very much a win-win situation.

---

[*] In a normal situation, the improvements could have been implemented without doing any further study. In this case, numbers were needed.

A challenge to intuition

A further study helps make the point. In this analysis, a group of grocery packers were evaluated for both performance and metabolic expenditure. The height of the work surface was manipulated to create the changes in arm posture.[*]

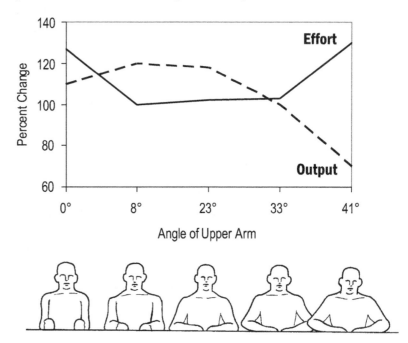

These results can be used to make statements on two levels that relate to the work-smarter-not-harder principle. First, on an elementary level, the graph tells us something about good working posture. When the elbows are held either too close to the body or too far away, (a) performance decreases and (b) the effort required for the job increases. The conclusion highlights the importance of working in neutral posture, that is, with your arms hanging naturally at your sides.

On a broader level, the results can be used to help us better understand productivity. Traditionally, we have tended to think that the harder we work, the more we produce. This study shows the opposite. Here, the easier the work is designed to be, the greater the output. It goes against our intuition.

Point

This study doesn't by any means prove that it is always true that working easier yields higher output. Rather, it only supports the claim that working in a neutral posture is better than working in an awkward or constrained posture.

But, it does challenge our tendency to believe that greater input always yields greater output. This is an important insight when we think about human well-being and the true concepts of productivity.

[*]Tichauer, E.R. *The Biomechanical Basis of Ergonomics.* John Wiley and Sons, 1978.

Good cliché

Speed is not synonymous with productivity. More input doesn't necessarily yield more output.

Productivity is best thought of as "working smarter, not harder." Although this phrase may be a cliché, the concept is valid and is worth repeating until everyone gets it.

Safety *and* efficiency

Productivity and efficiency are good things, if understood correctly, and can be achieved in ways that are compatible with employee well-being. Many professional ergonomists, myself included, often view themselves as efficiency experts as well as safety experts. Indeed, much of modern ergonomics is akin to old-fashioned methods engineering, often neglected in contemporary worklife. The tools and concepts of ergonomics can help resolve common conflicts and provide alternative directions that provide people and production.

People vs. machines

People are creative, can plan, and invent. We can react to unpredictable events, identify and choose options. We are good at interpreting data.

On the other hand, people are not so good at performing accurately in endlessly repetitive tasks. We are limited, wear out, are easily distracted, and inconsistent.

Machines are exactly the opposite.

Thus, we should design work so people can do what we are best at, and machines for what they are best at. Such statements can sometimes sound simplistic, but in practice, they are profound and provide good direction for resolving dilemmas in day-to-day decisions on the workplace floor.

**Take the toil
out of work**

As I have described thruout this book, at a certain level, this technique is very straightforward:

1. Look for things that are hard on people or unfriendly and frustrating. These are red flag alerts that the tasks are wasteful.
2. Study how people do these tasks step by step and then brainstorm ways to remove the physical demands. If someone says, "This is the way we've always done it," then look even more closely.
3. The headcount will more or less take care of itself.

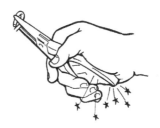

Work that hurts can be painfully slow.

**Final note:
Efficiency isn't
everything**

Paradoxically, there are times when thinking in terms of efficiency and productivity is precisely the wrong thing to do. Finding a more efficient way to build horse carriages was no way to compete against the Model T.

Instead, *effectiveness* is the answer in these situations and almost always trumps efficiency. A common wordplay among organizational enhancement experts is "Doing the *right thing*" (that is, having a good product or service) vs. "Doing the *thing right*" (that is, doing the work efficiently).

The maxim sometimes used in industrial ergonomics to make sure that effectiveness is taken into account is: *Don't fix anything that you shouldn't be doing in the first place.*

Bending and reaching
is a waste of time

Chapter 3
Understanding Musculoskeletal Disorders (MSDs)

One of the many important goals of ergonomics is to prevent MSDs, a class of disorders that basically amount to *wear and tear* on the tissue surrounding your joints. Every joint in the body can potentially be affected, but the lower back and upper limbs are the areas of most concern.

MSDs are actually quite common. Most of us will experience an MSD of one sort or another in our lives, often sports related or just lower back pain from everyday life. Most of the time, the symptoms are mild and disappear with rest. However, sometimes MSDs can become disabling.

The fact that MSDs occur should not be a big surprise. When machines do heavy or repetitive work, we expect fatigue and failure of moving parts, especially with poor maintenance. In many ways, humans are no different. Any "moving part" in the human can also fatigue and fail, especially if there is insufficient maintenance. MSD is simply the term we give to this human fatigue and failure.

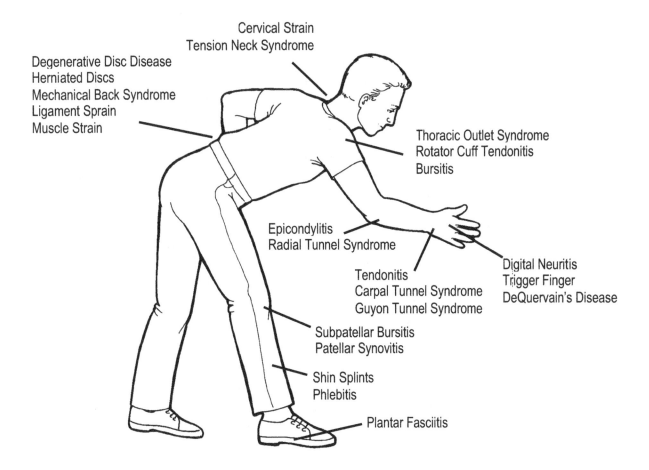

Cervical Strain
Tension Neck Syndrome

Degenerative Disc Disease
Herniated Discs
Mechanical Back Syndrome
Ligament Sprain
Muscle Strain

Thoracic Outlet Syndrome
Rotator Cuff Tendonitis
Bursitis

Epicondylitis
Radial Tunnel Syndrome

Digital Neuritis
Trigger Finger
DeQuervain's Disease

Tendonitis
Carpal Tunnel Syndrome
Guyon Tunnel Syndrome

Subpatellar Bursitis
Patellar Synovitis

Shin Splints
Phlebitis

Plantar Fasciitis

Synonyms

The term "MSD" is a catch-all category that lumps together a number of various disorders. Some medical experts are uncomfortable with this loose terminology, but from the point of view of workplace safety, it is a useful construct. Synonyms include:

- Cumulative trauma disorders (CTDs)
- Repetitive strain injuries (RSIs)
- Occupational overuse syndrome

Symptoms

The symptoms are often vague, slow in developing, and difficult to diagnose objectively:

- Soreness, pain, or discomfort
- Numbness
- Tingling sensations ("pins and needles")
- Weakness and clumsiness
- "Burning" sensations
- Limited range of motion
- Stiffness in joints
- Popping and cracking in the joints
- Redness, swelling, and local skin warmth

Treatment

In early stages, MSDs can in general be treated with ice packs and anti-inflammatory drugs. If the disorders progress, cortisone can sometimes help, but ultimately surgery may be needed. Some years ago, surgery was the treatment of choice, but now medical experts focus on conservative treatment, that is, ice packs, anti-inflammatories, and rest.

Wrist splints and other types of supports can provide relief in some cases. These devices are especially effective off-the-job, particularly during sleep. However, they generally should not be used as a preventative measure, since under some conditions, these devices can make problems worse.

Importance of early recognition

MSDs are common and as long as humans perform manual tasks, there will be a risk of suffering an MSD. However, it *is* possible to keep little problems from becoming big problems. This understanding has led to a strategy for medical intervention in the workplace to help insure that employees with symptoms are recognized at early stages and treated.

Prevention

By applying the principles and process from this book, the incidence and severity of MSDs can be minimized. The experience of many employers in the past few decades shows that prevention is possible if based on systematic efforts to:

- Identify problem tasks and make necessary ergonomic improvements.
- Identify people who are experiencing problems at early stages when treatment is easier and more successful.

MSD Rate — U.S. General Industry*

The following graph shows the trend for rates of recorded upper limb MSDs in U.S. general industry. The rates are expressed as the number of disorders per 10,000 employees. As you can see, the rates rose rapidly starting in about 1986 and continued to rise for nearly a decade, then after 1994 started to drop.

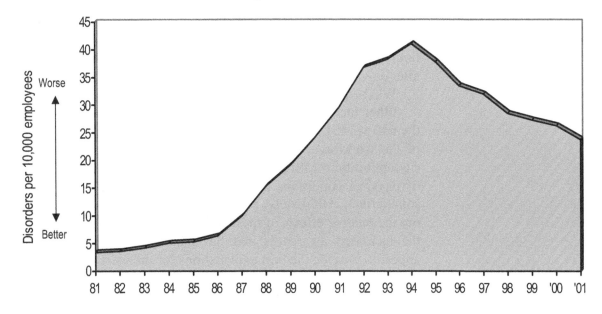

U.S. Bureau of Labor Statistics, 2005

Why the sudden increase?

MSDs have appeared on the scene for two main reasons: increased awareness and some subtle changes in technology.

Increased awareness — The awareness is related to newspaper articles on the problem, training classes, and even kits like this. People learn and recognize that they, too, are experiencing the same problems.

From all indications, MSDs have existed for hundreds, if not thousands, of years in every culture and part of the world. What is different today is merely that people are now reporting the problems.

Technological changes — The technological changes can best be explained by the example of the differences between the manual typewriter and the electronic keyboard. The former involved many types of motions: putting an individual sheet of paper into the roller, turning the knobs to move the paper into place, hitting the carriage return, stopping when you made a mistake to dab on some correction fluid, etc. The electronic keyboard involves using the same finger motions over and over. Moreover, more people use computers and use them for longer hours than typewriters. So the exposure level has also changed.

---

* See comments on p. 79. These rates are "repeated trauma" under the old system.

This change in technology has involved countless tasks in every industry. Where people used to perform tasks that involved a variety of different motions, now they perform tasks that require fewer types of motions, but more repetitively.

This subtle change has pervaded our society. For example, think of buying chicken in the grocery store. In the old days, the routine was to buy the chicken whole and cut it up yourself. The task was spread among the entire population. Now when you buy chicken, you have the choice of precut, boned, and/or skinned — which means there are workplaces where people do nothing but cut, debone, and skin chicken eight hours a day, day after day, and month after month.

What was once sporadic, is now repetitive.

Other factors — There may be additional factors related to the rise of MSDs. One is the aging workforce. A larger percent of the workforce is older than ever before, and age increases susceptibility to MSDs. Also, as societies, we are not in as good of physical conditions as we used to be, which probably has some effect. Ultimately, research studies should be able to sort out the relative effects of these different factors, but at this point, the increased awareness and the subtle technological change appear to provide the best explanations for the increased rates.

Sports analogy

A good reference point for understanding MSDs is professional sports. Ailments such as rotator cuff tendonitis among professional baseball pitchers are not uncommon, as can been seen in the sports pages of the newspapers. Baseball pitchers are young and in excellent shape. They warm up before pitching and do everything right. Yet they still have MSDs of the shoulder and arm because they perform repetitive, forceful motions using an awkward arm posture. It's too demanding on the human body to make that kind of motion continuously.

The same thing can be said of marathon runners. They are mostly young and in good shape and they warm up properly. Yet MSDs of the legs (shin splints, knee problems, ankle problems, etc.) are very common, perhaps universal among long distance runners.

Why the recent decrease?

The graph on the previous page visibly shows that MSDs have been decreasing since 1994. Ergonomics and medical management programs have begun to have impact. The effect is especially true in those industries that have developed massive ergonomics programs to combat their very high rates of MSDs. Examples are meatpacking and auto assembly where good ergonomics has clearly had a positive effect.

Glossary of Musculoskeletal Disorders

Most people do not need to know all the medical terms for MSDs, other than medical staff and perhaps some safety and human resources personnel. However, it can be helpful to recognize those terms that you might run across.

Hand and wrist

Carpal Tunnel Syndrome — Compression of the median nerve as it passes thru the carpal tunnel.

Tendonitis — Inflammation of a tendon.

DeQuervains Disease — Tendonitis of the thumb, typically at the base of the thumb.

Digital Neuritis — Inflammation of the nerves in the fingers caused by repeated contact or continuous pressure.

Ganglion Cyst — "Bible bumps" – Synovitis of tendons of the back of the hand causing a lump under the skin.

Guyon Tunnel Syndrome — Compression of the ulnar nerve as it passes thru the Guyon tunnel.

Synovitis — Inflammation of a tendon sheath.

Trigger Finger — Tendonitis of the finger, typically locking the tendon in its sheath causing a snapping, jerking movement.

Elbow and shoulder

Bursitis — Inflammation of the bursa (small pockets of fluid in the shoulder and elbow which help the tendons glide).

Epicondylitis — "Tennis elbow" – tendonitis of the elbow.

Radial Tunnel Syndrome — Compression of the radial nerve in the forearm.

Rotator Cuff Tendonitis — Tendonitis in the shoulder.

Thoracic Outlet Syndrome — Compression of nerves and blood vessels between the neck and shoulder.

Back, neck, and torso

Degenerative Disc Disease — Chronic degeneration, narrowing, and hardening of a spinal disc, typically with cracking of the disc surface.

Herniated Disc — Rupturing or bulging out of a spinal disc.

Ligament Sprain — Tearing or stretching of a ligament (the fibrous connective tissue that helps support bones).

Mechanical Back Syndrome — Degeneration of the spinal facet joints (part of the vertebrae).

Muscle Strain — Overuse or pulling a muscle.

Posture Strain — Chronic stretching or overuse of neck muscles or related soft tissue.

Tension Neck Syndrome — Neck soreness, mostly related to static loading or tenseness of neck muscles.

Hernia (various types) — A bulging of soft tissue thru an adjacent weak spot, often because of overexertion

Legs

Patellar Synovitis — "Water on the knee" – inflammation of the synovial tissues deep in the knee.

Phlebitis — Varicose veins and related blood vessel disorders (from constant standing).

Plantar Fasciitis — Inflammation of fascia (thick connective tissue) in the arch of the foot.

Shin Splints — Micro-tears and inflammation of muscle away from the shinbone.

Sub-patellar Bursitis — "Clergyman's Knee" – inflammation of patellar bursa.

Trochanteric Bursitis — Inflammation of the bursa at the hip (from constant standing or bearing heavy weights).

More information

Additional technical information and reviews of studies that show work-relatedness are available:

Putz-Anderson, V., *Cumulative Trauma Disorders: A Manual for Musculoskeletal Disorders of the Upper Limb*, Taylor and Francis, New York, 1988.

Panel on Musculoskeletal Disorders and the Workplace, Commission on Behavioral and Social Sciences and Education, National Research Council, *Musculoskeletal Disorders and the Workplace: Low Back and Upper Extremities*, National Academies Press, Washington D.C., 2001. This book is also available in PDF format for downloading from: http://books.nap.edu/catalog/10032.html

NIOSH, *Musculoskeletal Disorders and Workplace Factors: A Critical Review of Epidemiologic Evidence for Work-Related Musculoskeletal Disorders of the Neck, Upper Extremity, and Low Back*, DHHS (NIOSH) Publication No. 97-141, 1997. These materials along with a searchable database are also available online at www.cdc.gov/niosh/topics/ergonomics.

MSD Risk Factors

There are several factors that can increase the risk of MSDs, whether of the lower back, the arms, or the legs. The more factors involved and the greater the exposure to each, the higher the chance of developing a disorder. Note in particular that repetition is only one of several risk factors.

Tasks

Awkward postures — Body positions that deviate from neutral.

Static load — Using the same muscles for a period of time without change.

Pressure points — Direct pressure (or "contact stress") against any vulnerable part of the body.

Repetition — The number of motions made per day by a particular body part.

Force — The exertion required to make these motions.

Environmental — Exposure to temperature extremes or vibration and shock.

Work organization — Certain stressful situations related to organizational and administrative systems.

Personal issues

Physical condition — Poor personal fitness can play a role in the development of some MSDs

Diseases and conditions — There are a number of other diseases and disorders that can make the body more vulnerable to MSDs.

Considerations

Not all persons exposed to these factors will necessarily be affected, at least at the same level of exposure. That is why the word "risk" is used — the risk rises, but having a disorder is not inevitable.

There is sufficient scientific proof that indicates these factors increase the risk of MSDs. For the lower back, enough studies have been performed thru the years to be able to provide quantitative guidance on permissible levels of exposure (how many motions, at what levels of force) that can trigger a disorder. However, for the upper and lower limbs, these numerical guidelines are not yet known. All in all, the interest in disorders like carpal tunnel syndrome and tendonitis are new. The studies to develop guidelines for the hands, arms, and legs are taking place now and the results should be available in the next few years.

Prevention

In the meantime, enough information is known to support efforts to proceed along the lines described in this book.

The Arms and Legs

Although there are a number of specific MSDs for the arms and legs, most fall into two main categories: (1) tendonitis, and (2) nerve compression.

Tendonitis

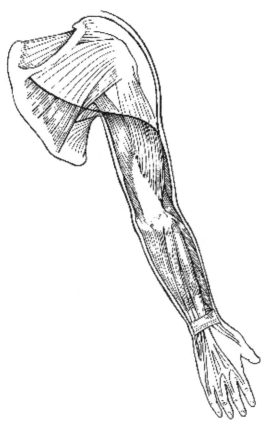

Tendons are found throughout the body, serving as links that connect muscle to bone. The tendons come into play every time a muscle is used to perform a motion involving bone structure.

Every time there are motions of this type, the tendons slide back and forth. As with any other moving part, with enough movement there can be wear and tear. When this damage occurs to tendons there is swelling and pain, and we call it "tendonitis."

The overall effect is like pulling a rope over a pulley. If the rope is pulled a few times, not much happens; but, if the rope is pulled thousands of times — or even a moderate number of times but with a heavy load — the rope can begin to wear.

Tendonitis can obviously occur in any locations where there are tendons. In the upper limbs, this usually involves:

- The tendons in the wrist — you can see these slide back and forth under your skin if you look at your wrist palm side up while opening and closing your fingers.

- The tendons in the back of your hand, in particular the one connected to the thumb — you can easily see these tendons by tensing your hand in a "claw" position.

- The tendons in your elbow — you can readily feel the one that connects your forearm with your bicep.

- The tendons in your shoulder — these are difficult to see or feel, but you can read about them in the sports pages during baseball season.

Nerve compression

Three major nerves run the length of the hand and arm, from the spinal column to the fingers. At several points it is possible for the nerves to be compressed. Sometimes making certain awkward motions or assuming certain postures can cause the pinching. Other times the swelling of nearby tendons causes the com-

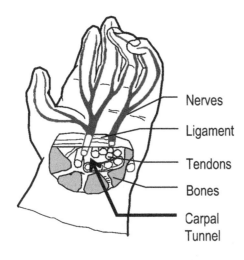

Nerves

Ligament

Tendons

Bones

Carpal Tunnel

pression. Another possibility is that factors such as awkward postures and static load reduce blood flow to the nerves, causing damage.

A common example of nerve compression is *carpal tunnel syndrome.* Inside the wrist is a small channel or tunnel—the carpal tunnel (*carpal* is the Latin name for "wrist"). One side of the tunnel is formed by a group of bones on the back of the hand. The other side is formed by a band of ligaments across the base of the palm. Inside the channel runs a major nerve and a number of tendons. If this nerve is injured either by swelling of nearby tissue or by reduced flow of blood, it can result in a kind of paralysis, with numbness and tingling sensations in the fingers. This is carpal tunnel syndrome, for which the classical symptom is "pins and needles" in the hands, especially at night.

This type of compression can also occur in various other parts of the arm, such as in the shoulder and elbow. Symptoms tend to be the same, that is, "pins and needles" in the hands.

Reynaud's phenomenon is another type of cumulative disorder that is sometimes included in workplace ergonomics. Altho the condition can be caused by other factors, in the workplace the concern is for heavy vibration, such as from jack hammers. The vibration can cause the blood vessels in the hand to spasm and constrict the flow of blood, causing the fingers to feel cold and sometimes blanch.

The lower extremities

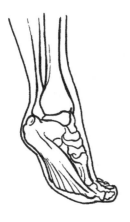

In general, the same types of problems that occur in the hands and arms can occur in the legs and feet, that is, various types of nerve compression and tendonitis. There are some differences, however, between the upper and the lower limbs. One difference is in the relative importance of certain risk factors.

For example, with the feet and legs, contact stress is much more of a factor than with the hands and arms. The primary example is plantar fasciitis ("heel spurs"), which stems from repeated standing, walking, or running on hard surfaces. Similarly, the knees can be damaged from constant kneeling, such as with carpet layers or gardeners.

The Back and Neck

The back is also vulnerable to cumulative trauma. We often think of a back injury as caused by a single event, like lifting a particularly heavy load. However, more than likely, the injury is the result of the cumulative effect of bending, twisting, or excessive sitting or standing. A single event may trigger the injury, but it may be merely the "straw that broke the camel's back."

The vulnerable discs

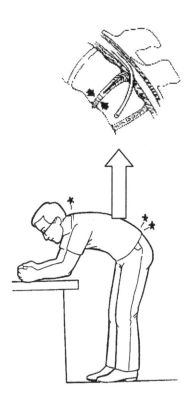

The spinal column consists primarily of bones — the vertebrae — separated by pliable discs. Thru the center of this column run the nerves of the spinal cord.

The discs provide cushioning between the vertebrae. They enable us to bend and twist, and have the incredible range of motion that our backs have. Unfortunately, with enough bending and twisting, especially while carrying a load, the discs can suffer wear and tear. After a time the discs can narrow, harden, and the surfaces fissure and crack; what we call "degenerative disc disease." Once weakened, the discs can bulge out, strain, or herniate.

There are a variety of other ways that the back can be injured by overuse or overexertion. Examples are sprains of the ligaments, strains to the muscles, and a condition known as mechanical back syndrome.

The neck is a part of the back, and injury there can occur in the same manner as anywhere else in the spinal column. The most commonly injured part of the back, however, is the lumbar region.

The jelly donut effect

The discs have somewhat the construction of a jelly donut — or maybe a three-week-old jelly donut. There is a tough but flexible surface on the outside and a kind of "goo" on the inside (to use a highly technical medical term).

The effect of bending is something like the following. When you squeeze a jelly donut on one side really hard, there is a risk that the jelly will shoot out the other end. Similarly, when you bend over, you place pressure on one side of the disc, which increases the risk for the "goo" to herniate out to the other side.

If the bulging of the disk is severe enough, it can press against the nerve and you can feel severe pain, including burning leg sensations.

Time Sequence — Onset of Arm and Wrist MSDs

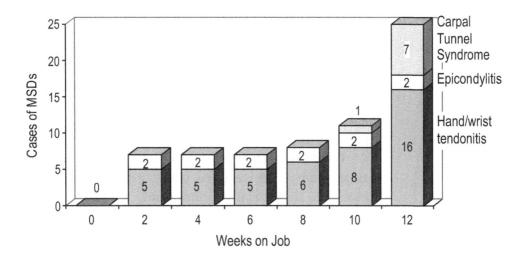

This graph shows the time sequence for the onset of upper limb MSDs. This study[*] tracked the experience of 40 new hires in an operation that involved highly repetitive tasks over a 12-week period in an assembly operation. (A control group, not shown here, had dramatically fewer problems.)

Hand/wrist tendonitis occurred two weeks after trainees started the job. Cases increased gradually until the 10th and 12th weeks when the cases sharply increased, including the first cases of carpal tunnel syndrome.

This sequence may be common, but it is possible for development of MSDs to occur in both shorter and longer periods.

Not New

MSDs are neither a new phenomenon nor attributable to the modern mechanized workplace. Once you become aware that these disorders could have occurred in the past, you start seeing references in novels and history books to what would undoubtedly be characterized today as MSDs. Even in older movies and television shows characters talk about their "rheumatism" acting up, which could well be the old vernacular for MSDs. The only thing particularly new about these disorders is that they are being reported in modern society where health and well-being receive more consideration than in the past.

The following pages provide examples from the classic book of occupational medicine as well as from a number of novels and other historical sources.

---

[*] Adapted from Tichauer, E.R.. *The Biomechanical Basis of Ergonomics.* John Wiley and Sons, 1978.

MSDs in the 1700s

An Italian physician, Dr. Bernardino Ramazzini (considered the founder of occupational medicine) wrote a book in 1713 that clearly describes maladies that today would be termed MSDs. Most of the book *Diseases of Workers* contains descriptions of occupational diseases including items such as lung disease among miners, lead poisoning among potters, and even overtaxed minds among "learned men." But amidst all these disorders are descriptions of ". . . the harvest of diseases reaped by certain workers . . . [from] irregular motions in unnatural postures of the body."

Writing with a quill pen — "The maladies that affect the clerks arise from three causes: first, constant sitting; secondly, incessant movement of the hand and always in the same direction; and thirdly, the strain on the mind . . ."

"The incessant driving of the pen over paper causes intense fatigue of the hand and the whole arm because of the continuous . . . strain on the muscles and tendons."

"An acquaintance of mine, a notary by profession, who, by perpetual writing, began first to complain of an excessive weariness of his whole right arm which could be removed by no medicines, and which was at last succeeded by a perfect palsy of the whole arm. . . . He learned to write with his left hand, which was soon thereafter seized with the same disorder."

Standing — "Those who work standing . . . carpenters, sawyers, carvers, blacksmiths, masons . . . are liable to varicose veins . . . [because] the strain on the muscles is such that the circulation of the blood is retarded."

"Standing even for a short time proves exhausting compared with walking and running though it be for a long time . . . . Nature delights and is restored by alternating and varied actions."

General — "[I have seen] workers in whom certain morbid affections gradually arise from some particular posture of the limbs or unnatural movements of the body called for while they work. Such are the workers who all day stand or sit, stoop or are bent double, who run or ride or exercise their bodies in all sorts of [excess] ways."

Sitting — "Those who sit at their work suffer from their own particular diseases." [As noted back in Roman times by the learned slave] Plautus, 'sitting hurts your loins, staring, your eyes.'"

Repetitive hand motions — "I have noticed bakers with swelled hands, and painful, too; in fact the hands of all such workers become much thickened by the constant pressure of kneading the dough."

MSDs in Literature

Swelled wrists and rheumatic backs — From Herman Melville's *Moby Dick,* published 1851 about the life he experienced in whaling ships in the early 1800s:

". . . they have swelled their wrists with all day rowing . . ."

". . . a roll of flannel for the small of someone's rheumatic back . . ." [i.e., a lumbar cushion].

Rheumatism in her arms — Mark Twain, *Pudd'nhead Wilson,* written in 1893 about life in the early 1800s:

"She got a [job] as second chambermaid on a Cincinnati boat in the New Orleans trade. During eight years she served three parts of the year on that boat, and the winters on a Vicksburg packet. But now for two months she had had rheumatism in her arms, and was obliged to let the wash tub go. So she resigned."

"She's kind of crippled in the arms and can't work."

Rheumatism ointment — Willa Cather, *Shadows on the Rock,* written 1931:

"Madame Renaude says she could never milk her cows in the morning if she did not put rheumatism ointment on her hands at night."

Cowboy hanging pole — John McFee, *Rising from the Plains,* referring to the 1910 era:

"Many [cowboys] were already stooped from chronic saddle-weariness . . . and spinal injuries that required a "hanging pole" in the bunkhouse. This was a horizontal bar from which the cowboys would hang by their hands for 5-10 minutes to relieve pressure on ruptured spinal disks . . . ."

The Wayward Desk — Herman Melville, *Bartleby the Scribner,* set in the early 1800s:

"Nippers could never get his table to suit him. He put chips under it, blocks of various sorts, bits of pasteboard, and at last went so far as to attempt an exquisite adjustment by final pieces of folded blotting paper. But no invention would answer. If, for the sake of easing his back, he brought the table lid at a sharp angle well up towards his chin, and wrote there like a man using the steep roof of a Dutch house for his desk — then he declared that it stopped the circulation in his arms. If now he lowered the table to his waistbands, and stooped over it in writing, then there was a sore aching in his back."

Planting Rice — From V. S. Naipaul, *A House for Mr. Biswas,* published 1961 about life in Trinidad in the 1930s:

"Owad said, 'You know the labour that it is to plant rice. Bending down, up to your knees in muddy water, sun blazing, day in day out.'

" 'The backache,' the widow said, arching her back and

putting her hand where she ached. 'You don't have to tell me. Just planting that one acre, and I feel like going to the hospital.' "

More Rice — From Jung Chang, *Wild Swans*, 1991, about life in China in the 1970s:

"I preferred transplanting rice shoots. This was considered a hard job because one had to bend so much. Often at the end of the day, even the toughest men complained about not being able to stand up straight."

That's Why it Hurts — from John Steinbeck's *East of Eden*, published in 1952 about life in the late 1800s:

"One day and Samuel strained his back lifting a bale of hay, and it hurt his feelings more than his back, and for he could not imagine a life in which [he] was not privileged to lift a bale of hay. He felt insulted by his back, almost as he would have been if one of his children had been dishonest.

In King City, Dr. Tilson felt him over. The Dr. grew more testy with his overworked years.

'You sprained your back.'

'That I did' said Samuel.

'And you drove all that way in to have me tell you that you sprained your back and charge you two dollars?'

'Here's your two dollars.'

'And you want to know what to do about it?'

'Sure I do.'

'Don't sprain it anymore. Now take your money back. You're not a fool, Samuel, unless you're getting childish.'

'But it hurts.'

'Of course it hurts. How would you know it was strained if it didn't?' "

Marine Sergeant — from S.L.A Marshall, military historian, referring to the Korean War in 1951:

"I found a note from my clerk-typist, Sergeant Shafer, informing me that he had been so badly overworked in the field that he was turning into the hospital with a lame wrist."

Butcher's Wrist — related to me by a retired meatpacking supervisor:

"Yeah, I remember my old man, a ham boner by trade [in the 1920s]. His hands would swell up like a prize fighter."

Cornhusker's Disease — related to me by a meatpacking superintendent:

"When I first hired on as a kid [in the 1930s] and started boning, my hands swelled up and they hurt. I went to the doctor and he diagnosed me as having cornhusker's disease."

MSD Rates by Industry and Occupation

**Injury rates,
old and new**

In 2002, the U.S. Bureau of Labor Statistics (BLS) launched a major change in its system for reporting on MSDs. It also changed its industry classifications in 2003 to account the growth of service industries in recent years. These changes provide a more accurate depiction and are much more equivalent to the definitions used in this book.

Unfortunately, the changes mean that the data prior to 2002 are not comparable with data after 2003. On the other hand, the comparisons between the old and new shed interesting light on risk factors in different industries.

The old system was based on an injury classification called "repeated trauma," which consisted primarily of hand and arm MSDs like tendonitis and carpal tunnel syndrome. The new system includes back pain, strains, sprains, and other over-exertion injuries along with the classic repetitive motion disorders like tendonitis and carpal tunnel syndrome.

The old system was also based on the Standard Industrial Code (SIC) classification. The new system involves the North American Industrial Classification System (NAICS), which was developed to reflect changes in the service industry and to standardize the data reporting with Canada and Mexico.

**Manufacturing:
hand and arm MSDs**

The change in how injuries are classified provides important insights on the physical risk factors in different industries. The old system captured primarily hand and arm MSDs, which are most prevalent in manufacturing establishments, as shown below, when the injuries are analyzed by major industry sectors.

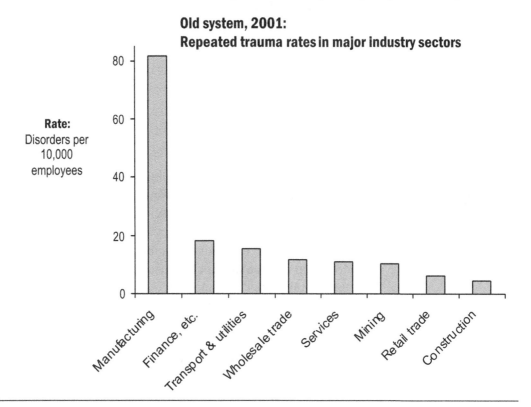

**Old system, 2001:
Repeated trauma rates in major industry sectors**

Rate: Disorders per 10,000 employees

**Meat, poultry,
auto, and sewing**

A more detailed breakdown of establishments by four-digit SIC code shows that "repeated trauma" is highest in the meat and poultry industries, the auto industry, household appliance manufacture, and certain apparel operations. These industries have consistently been at the top of the list over the past decade. In particular, meat, poultry, and the auto industry have continually been in the top three or four. Other types of apparel manufacture have sometimes been higher than men's/boy's trousers and shoes, but one sewing operation or another has consistently been high. Similarly, other types of fabricated metal manufacture have been higher than household appliances, but there has always been some type of metal fabrication and assembly at the top of the list over the past decade.

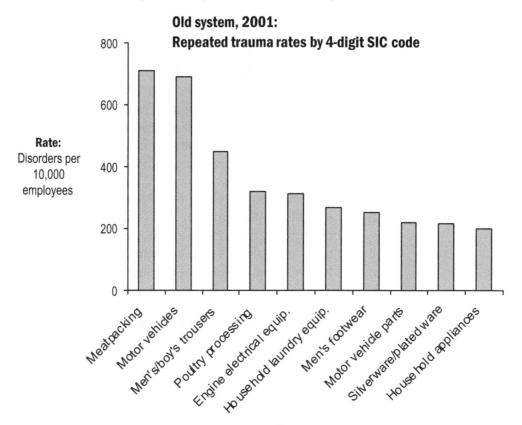

**Old system, 2001:
Repeated trauma rates by 4-digit SIC code**

Rate: Disorders per 10,000 employees

**Service industry:
back injuries**

The pattern changes completely once lower back pain and various types of strains and sprains are included. As mentioned above, the new BLS system uses a category called MSD that is more inclusive As shown on the opposite page, under the new system, the service industry emerges as the prime source of MSDs, in particular, operations with a lot of repetitive lifting: baggage and package handling, beverage bottling and delivery, driving, and warehousing.

The automobile industry is still in the top 10, but meat and poultry processing is nowhere to be seen (ranked 136rd on this list). Nor is there any type of sewing operation.

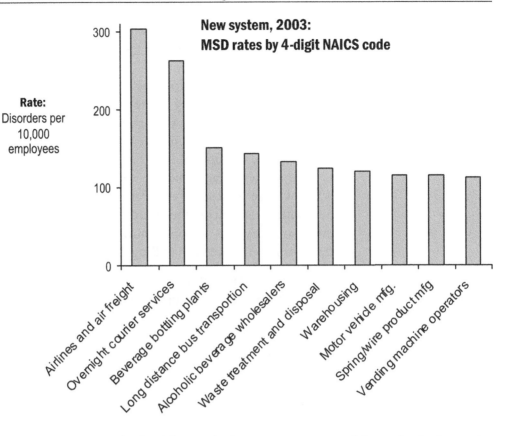

New system, 2003:
MSD rates by 4-digit NAICS code

Health care

The health care industry warrants special emphasis due to its huge size. When we look at the sheer number of injuries (unlike the above graphs, which are all *rates*, that is, injuries per 10,000 employees in the industry), the problem clearly emerges, as shown below.

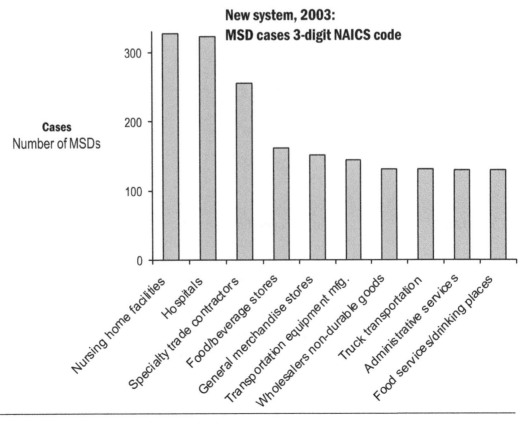

New system, 2003:
MSD cases 3-digit NAICS code

Computer work vs. physical labor

Also, note that wrist disorders among computer users do not appear on these charts in any fashion. One reason again has to do with *cases* versus *rates*. The number of cases of MSDs among computer users may be high, but the number of people using computers is vastly higher, so the *rate* is not as high as in assembly type jobs. A more important reason is that the analyses above are for *industry* and computer users are spread thruout many categories.

We see a different picture when we look at *occupation* and when we use total *number of injuries* rather than *rates*. Office occupations appear as the fourth highest on the list. However, even so, they are a poor fourth compared to occupations in transportation, material handling, and production.

New system, 2003:
MSD cases by Standard Occupational Classification (SOC)

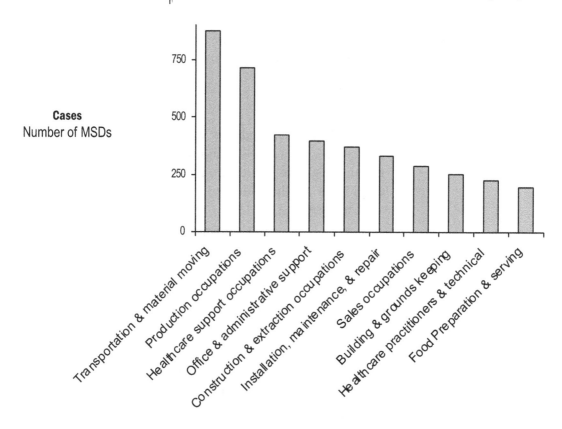

Cases
Number of MSDs

BLS updates

More recent data are not available at the time of this printing. However, the Bureau of Labor Statistics (BLS) maintains an excellent website where the most recent occupational injury, illness, and fatality data are easily accessible.

New focus on the service industry

Undoubtedly, the change in report from "repeated trauma" to the broader "MSDs" will lead to greater emphasis on the service industry. Manufacturing still has its problems, but as everything else, the service industry will be the focus of the future.

Chapter 4
Case Examples

Many of the world's largest and best-known companies have invested heavily in ergonomics in recent years. Good examples are companies in the automobile and computer products industries. In many facilities in these industries, ergonomics has become a part of everyday business.

However, some of the best company-wide financial data come from smaller enterprises. The reason is that in large operations, like an auto company or even in a single auto assembly plant, there are so many variables that often it is difficult to sort out which changes are related to which financial savings. In smaller operations, the causes and effects can be much more visible.

This chapter contains several case examples from smaller companies that didn't institute any change other than starting a formal ergonomics process. Thus, the situations were a kind of natural experiment in which only one variable was changed. Additionally, these case examples involved task improvements that are especially noteworthy.

These examples demonstrate:

- The bottom line benefits of good ergonomics
- A description of how the process worked in a particular case
- A sampling of specific improvements that provide a sense of the changes that were made in problem tasks

Case Example 1:
Ergonomics Saved a Failing Company

In some ways, this example is unfair, since the workers' compensation costs in this company were horrendously high. Yet, these situations do exist, and undoubtedly companies fail every year because of the absence of good ergonomics. Also, this story involves an especially creative solution.

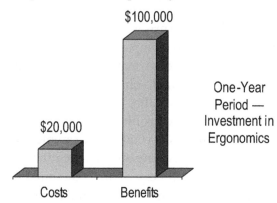

500% ROI due to immediate drop in workers' compensation costs.

Workers' comp insurance cancelled

This case involves a small machining operation with 60 employees in a small town in the rural Midwest. The plant had experienced so many musculoskeletal disorders (MSDs or "wear-and-tear" injuries) that their workers' compensation insurance carrier canceled their coverage and the plant was forced to enter a state pool to obtain insurance. As the plant manager said, "It's a little like getting a DWI. Yes, you can get insurance, but, boy, do you have to pay for it."

In time, these serious workers' compensation problems threatened the viability of the company. Fortunately, the ergonomics intervention stopped the financial hemorrhage and rescued the business.

The company manufactured metal parts used in mainframe computers, taking advantage of the company's capabilities for high tolerance machining. The plant consisted of two departments: (1) the machine shop, filled with millions of dollars of high-end machine tools, and (2) the deburring area, crammed with people standing at workbenches. Both departments were affected, but the deburring area was the primary problem.

Organization

The process in this company was very modest and informal. There were no new committees, since the issue was a dominant one for the company and normal line management and staff handled all the issues as part of their normal duties. Since the plant was small, it was easy to have impromptu meetings with employees either in the lunch room or on the work floor. There were several manufacturing engineers on-site, plus a facilities engineer from the parent company located in a large city about

90 miles away. I spent an initial two days in the facility, then returned one day per month for the following four months. The facilities engineer spent one or two days per week on site.

Training

One of the first steps taken was to hold a 45-minute meeting of the deburring employees to introduce me to the group, explain the project, and ask for their cooperation in finding improvements. I gave a presentation on the principles of ergonomics and musculoskeletal disorders, then the employees were given a chance to talk about problems they were having. It became clear as a result of this meeting that a larger percentage of the employees were experiencing problems — about 50% — than had ever reported any difficulties to management. Furthermore, the employees pointed out the task which they considered the most difficult: deburring the "slotted pole."

Task evaluation

The facilities engineer and I reviewed the slotted pole task and discussed specific issues with employees in the area and the supervisor. The slotted pole was a cylinder about five inches long mounted on a stubby base. The inside of the cylinder was somewhat hollow, with three slots trisecting most of its length, hence its name. The job was to take the parts that had been recently machined and remove burrs and sharp edges. Because it was to go into a mainframe computer, all slivers of metal had to be removed completely and the rough surfaces polished. Quality requirements were particularly high, since the slightest fleck of metal could ruin an entire computer.

The task of deburring the slotted pole was unpopular. Consequently a variety of people had been trained to do the work and rotated through the job to spread the burden equally.

One segment of the task consisted of holding the pole in the left hand and manipulating it, while the right hand performed various motions such as sanding its edges and surfaces. The sandpaper was held in an awkward "pinch" grip with fingers extended to reach the inside of the cylinder, thus creating a high force on the right-hand wrist. However, the employees stated that the worst part of the task was with the *left* hand, caused by the continuous gripping ("static load") and constant manipulation of the part.

Based on this simple assessment (that is, no measurements or numbers) the facilities engineer, the supervisor, a couple of employees, and I started brainstorming ideas for improvement. We stood around the workbench and tried to imagine alternatives. The first goal was to fixture the part to relieve the left hand from the constant gripping forces. The fixture needed to rotate, however, to accommodate the need to access the part in different orientations.

Brainstorming

Initially, we considered a simple "Lazy Susan" type fixture. However, a concern developed that the need to continuously rotate the fixture manually would merely transfer the strain from the hand to the shoulder. So we started exploring the idea of powering the fixture. This idea too was discarded, since the "slotted pole" needed to be both rotated at varying speeds then indexed at one-third turns. A machine to achieve both requirements would cost too much in money and development time.

Brainstorming then refocused back to the Lazy Susan, with the added idea of running a shaft down to the floor where it could be easily manipulated by the feet. As the facilities engineer began to sketch the needed device, it occurred to the group that a similar device had been invented long ago and was readily available.

Potter's wheel

The company purchased and installed a potter's wheel. Technology that was 4000 years old was used to solve a problem in a plant that made computer parts.

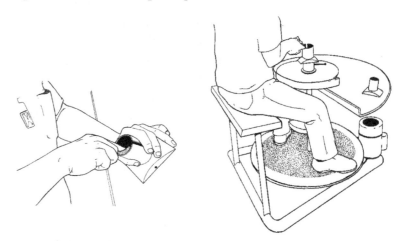

Before: Left-hand manipulation   After: The potter's wheel

There were many features of the potter's wheel that could be easily applied to this task. The plate that was normally used by potters to support clay was modified to support a fixture to hold the slotted pole. The shaft that connected the plate (and now the fixture) to the wheel beneath could be manipulated easily by the feet, both to rotate evenly at varying speeds and to index precisely.

This off-the-shelf device worked successfully and served as the basis for further developments. Specifically, the fixture eliminated all of the left hand motions and exertion in manipulating the slotted pole, thus, in one step, reducing the overall wrist stress by at least 50%.

Another part of the task involved the use of a generic "deburrer's knife." As shown in the illustration on the next page, this knife was used to reach deep into the center of the cylinder to scrape out burrs. The illustration also shows several

Deburrer's knife

other issues. One is the awkward postures of the wrist and elbow. Another issue is the grip on this generic device, which was so inadequate to the task that it was not used at all. Finally, the illustration once again shows the problem with the left hand, an extended pinch grip that was eliminated with the use of the fixture now mounted on the potter's wheel.

Deburrer's Knife   Custom, two-handed knife

Once the part was attached to the fixture, then the left hand became available to help manipulate tools. In this case, a special two-handled tool was designed specifically to reach into the slots to scrape out burrs. The tool could be inserted into the slot and pulled, providing three great benefits: (a) the tool placed the wrists and elbows in much better posture, (b) it divided the force between both hands and (c) it took advantage of the larger muscle groups in the upper arms and the shoulder and torso.

Cutting wasteful motions

A further problem in the original task was sanding the outside of the part. Considerably repetitive and rather forceful motions were required to complete the task. Once again, the left hand was used as a fixture in doing this step of the job.

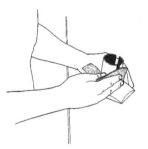
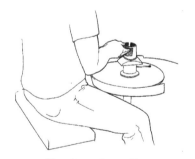

Hand sanding, before   Hand sanding, after

The potter's wheel that the facility purchased represented the high end of potter's wheel design, and came equipped with an electric motor to spin the wheel. As applied to this deburring task, the wheel could be manipulated by the feet in order to index and move the part slowly, but the motor could also be used to spin the wheel. In this mode, the employee could simply hold a piece of sandpaper rather loosely to the slotted pole, polishing it quickly and cleanly with little physical effort.

Continuous improvement

As experience was gained with the potter's wheel, a variety of modifications were made. The plate (to support the potter's clay) was removed, since it had no function anymore. The fixture was angled with the use of a gearbox to orient the slotted pole to the employee for improved access. A storage area that was part of the original equipment, but a bit out of reach, was raised up and moved closer. A task light was added.

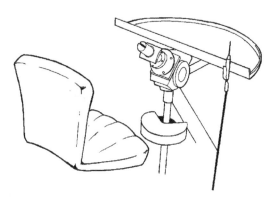

Modified potter's wheel

Creative maintenance man

Two additional items of special note also evolved. Both involved the plant maintenance man, who in the course of events had become actively engaged in redesigning the potter's wheel. He happened to be both a Harley-Davidson and a NASCAR enthusiast and had a supply of spare car and motorcycle parts available.

The original potter's stool had no back support and lacked cushioning for the seat. He replaced it with a car seat and wired it into the plant electrical system so that employees could adjust the height and leg distance by the touch of a button.

The other special feature involved a problem with the basic design of the original potter's wheel. Once the wheel was spinning, there was no way to stop it quickly. Potters had no particular need to stop the wheel as abruptly and frequently as did these deburring employees. Consequently, to stop the wheel, he attached a Harley-Davidson brake, complete with a large and chrome-plated lever that stood out conspicuously to one side.

Habits changed

On previous occasions, the supervisor had encouraged the employees to use various powered deburring tools that were available. However, the employees resisted, stating that the powered tools were awkward to use. As it turned out, once the employees began to use the potter's wheel, they also began to use the power tools successfully. The employees had become so used to doing the task in a certain way, that any change (even ones that were better) *seemed* awkward. However, once the entire task was disrupted and whole new techniques needed to be learned, the employees were able to incorporate the power tools.

Improving other tasks

After the initial changes were made on the slotted pole, attention shifted to other tasks, again involving employees and brainstorming improvements.

The other deburring tasks involved working on large, traditional work tables. Employees often worked hunched over these tables, with poor back and neck postures. The work benches were fixed-height and could not be adjusted for taller workers. Moreover, since two or more people typically worked at each table, the tables could not accommodate everyone easily, even if they had been adjustable. Additional issues included: no particular place provided for tools, occasional long reaches for materials, and the employees stood all day on the concrete floor.

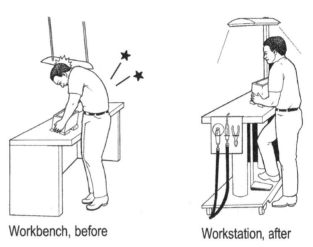

Workbench, before   Workstation, after

Workbench

The first decision was to procure sturdy, adjustable-height work tables. However, none could be found on the market when this case study occurred. Happily, once again creative thinking led to an off-the-shelf product.

The item purchased was a standard industrial die-cart — originally designed to lift heavy dies into presses. Altho it was intended for another purpose, it had many of the features that were needed and it was easier to modify than build a new workstation from scratch.

The die cart was modified in several ways, primarily by adding a heavy duty fixture to the top. Also, a one-quarter inch sheet of plastic was attached to the surface to provide more of an appearance of a work table and to protect tools and parts from the steel. Brackets were mounted to the sides of the cart to hold tools, fixtures and a task light.

Used as a work station, these carts could readily be raised and lowered to adjust for individual height. Furthermore, the die cart had several additional features which unintentionally came into play:

- The dimensions of the top work surface were small, only about three feet wide and two feet deep — quite sufficient for the task at hand. The unintended benefit was that once all employees were equipped with these tables and the old, larger tables discarded, considerable floor space was gained. A previously congested area became rather roomy.

- The die carts came with wheels, needed for their original application. These new work stations were thus mobile and could be used as the basis of a modular, flexible manufacturing system. The engineers created several work cell areas using pedestals attached to the floor and equipped with electric outlets and air couplings. Employees were able to wheel their workstations to the appropriate cell, hook up and work as a unit on a particular part. As production required, they shifted quickly from one cell to another.

- The base of the cart served well as a foot rest, whether sitting or standing.

- These individual workstations could be personalized for each employee to promote an individual's own identity.

Finally, the workstations were designed for standing height, and tall stools were provided to permit employees to sit or stand as they chose. The stools themselves were adjustable and had other ergonomic features. Anti-fatigue mats were provided for relief when standing.

Workstation built around a fixture

Both the potter's wheel and the above workstations illustrate a subtle, but important concept in workplace design. Rather than big, generic workbenches that take up a lot of space and are not adjustable or movable, it is better to think in terms of designing a workstation *around* a fixture.

This concept can yield workstations that are both more human compatible and focused on the specific product at hand, hence, more efficient. In addition, this approach generally yields workstations requiring less floor space.

Machining areas

Although the machining area was not the focus of initial efforts, eventually engineers identified ways to make improvements here as well. For example, an articulated arm, previously purchased for another project but found to be infeasible for that task, was successfully used to lift the heavy parts in and out of the machine tools. Additionally, an ingenious conveyor line was constructed to move parts from one machine tool to another instead of loading onto a cart, moving a few feet, then unloading the cart. Thus considerable time-wasting and stressful repetitive motions were eliminated.

Results

Many positive results emerged from these changes:

Productivity increased — Using the new workstation, employees were able to produce six parts per hour compared to the previous five. However, the work standard was not changed since the management's focus was on quality rather than quantity. Furthermore, the managers thought that continuing to produce only five per hour would permit additional rest time to further reduce the risk of injury.

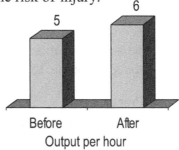

Output per hour

Rejects eliminated — With the workstation changes and the extra time to do the job, costly batch rejection rates were reduced to nearly zero.

Employee satisfaction increased — The employees clearly appreciated the new workstation. In particular, employees complained whenever the potter's wheel was taken out of service to be modified, forcing them to do the job the old way. An additional potter's wheel ultimately had to be purchased as a standby.

90% drop in claims — The expensive musculoskeletal injury cases were prevented and worker's compensation costs dropped an astounding 90% during the following year. Moreover, within *one year* the company was out of the state insurance pool and able to obtain normal insurance. This was the first time in the history of the state that a company was able to exit the pool in *less than three years*.

500% Return on Investment

The costs for improving the deburring area amounted to a one time investment of about $20,000, including both administrative time and purchase of equipment. The savings in worker's compensation during the following year alone were $100,000, that is, a 500% return in one year.

Note especially that the investments were one-time costs only, but the benefits continue to accrue year after year.

Postscript: Fixing the root cause

After the initial changes described above were implemented, the engineers recognized deburring as a problem in a way in which they had not previously. Ultimately, they found ways to improve machining capabilities to reduce the burrs.

Thus, they eliminated most of the need for deburring. They finally fixed the root problem.

Lessons

There are many lessons to be learned from this example. The primary message is that a creative workplace ergonomics process can simultaneously reduce injuries, cut costs, and increase productivity.

There are also many insights regarding ways to solve specific workplace problems: using fixtures, the importance of dedicated tools, and so forth. These strategies can be adapted for many industries, including service jobs and even office facilities.

Finally, there are many lessons that provide guidance in setting up an effective workplace process. These concepts will be addressed more systematically later in this book, but to drive the ideas home, the following points can be made:

Process

- It is vital to talk to employees in order to learn about the job and gain insights into problems and solutions. Some details can be gained in no other way.
- Often, the specific problems of a job can be identified relatively easily, and without the need for elaborate studies or measurements. Typically the difficulty is in finding solutions.
- The problem-solving process need not be elaborate or highly technical. Creative thinking is often more important.
- Everyone can make a contribution. Maintenance personnel, once provided with the principles of ergonomics and the goals of the program, can be especially gifted in making creative improvements.
- Incorporating change can be difficult for a variety of reasons, one of which is the chore of learning a highly refined work technique all over again. Sometimes it is easier to make a big change than a little one.

Solutions

- Anything which solves a problem is ergonomics — the solution does not need to be a device normally thought of as an "ergonomic" product.
- Unconventional, "hare-brained" ideas often lead to good results. Solutions can be found in many areas, even from equipment not normally associated with the industry in question.
- Reducing motions does not necessarily mean slowing down the job. On the contrary, with good design, the task can be completed faster, but with less manual work.
- Continuous improvement applies to ergonomics as much as to any other aspect of the workplace. There is no final ergonomic fix — there are always ways of improving equipment.
- One should never neglect the root cause of the problem.

Case Example 2
Ergonomics Helped a 200-Year-Old Firm Dominant in its Market

This company was cited by OSHA for repetitive motion injuries, paid a $300,000 fine, and was required to establish a formal ergonomics program. Fortunately, the company set up an excellent process that resulted in a $1 million savings over five years.

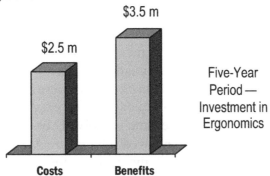

OSHA citation

The company was involved in making paper and high quality paper products, such as bank notes and currency paper. Its origins dated back to the 18th century and it had competed so successfully thru the years that it became dominant in its market niches, and in the case of one of its products, a legal monopoly.

Approximately 1200 people were employed in several facilities in New England. Several tasks in the company were very labor-intensive and involved highly repetitive motions of the arms, resulting in MSDs among a number of employees. The company began to address these injuries, but a subsequent OSHA inspection resulted in a citation and fine. The company responded positively by developing a more aggressive and comprehensive ergonomics effort.

Program description

Key aspects of the company's ergonomics program included the following:

Senior management added ergonomics as a regular agenda item in their ongoing meetings, which permitted them to monitor overall progress and help make strategic decisions. Special attention was given to accountability within the management system, to the extent of adding formal ergonomics factors and injury reduction into individual goal-setting and performance evaluations.

Ergonomics committees were created in each of the facilities and a capable supervisor assigned as company-wide ergonomics coordinator. Employees were involved at all levels, including serving on committees, videotaping, job analysis, administering surveys, auditing, recommending improvements, and monitoring progress. This was the first formal effort of the company in

Involvement

employee involvement and led to other ways in which employees were involved in activities traditionally reserved for managers.

The company retained me to provide training thruout the organization and help establish the overall process. As part of this process, the committees started evaluating *all* the tasks in their facilities, beginning with the sited jobs, but ultimately including *every* job performed in their operations. Initial task evaluations were non-quantitative in nature and focused on simple identification of ergonomics issues and brainstorming for possible improvements. As the program evolved, quantitative approaches to evaluating risk factors were introduced to verify in several cases that risk factors were being reduced.

The company upgraded its existing medical system. The on-site nurse became involved on the corporate ergonomics committee, reviewed jobs where MSDs had been reported, and became part of the process for recommending improvements. Employee education focused on early recognition and reporting of symptoms; the approach to treatment was conservative. A myotherapist was retained to provide early, hands-on treatment, and to teach stretching exercises. The company also instituted annual discomfort surveys, in part as a way to evaluate progress.

Of particular importance, the editor of the company newsletter started including ergonomics updates and success stories in each issue, keeping everyone updated and providing recognition to individuals and groups who were making especially good contributions. Additionally, the company celebrated milestones at various intervals (such as steak dinners for all employees), which promoted a sense of momentum. These steps were considered essential for changing mindsets in this old, established firm.

As the program evolved, many of its aspects changed, such as structure of committees, approaches to risk factor evaluation, how often meetings were held, and the relative focus between engineering and administrative controls.

Yankee ingenuity

The orientation of their ergonomics process deliberately promoted common sense and low-tech solutions. I started using the phrase "Yankee ingenuity" to make the field of ergonomics more accessible and within the traditions of these New Englanders. The use of the term as a synonym for ergonomics helped demystify the field and tapped into the workforce's heritage of innovation and creativity.

Workstation improvements

Most of the improvements involved standard ergonomics devices, such as vacuum-assisted hoists and lift tables, plus some automation. Several unique innovations were devised, however, which merit description.

Mechanical flipper

Task — In one of the company's facilities, about 40 of the employees operated small printing machines to imprint cards and envelopes for events such as weddings and births. The task involved using the left hand to feed the machine and the right one to remove the product, flip it over and place it on a small conveyor that passed the printed materials thru a small drying oven. Thus the employees moved their right arm and flipped their right wrist for each card and envelope printed.

There was nothing unusual in this set up. The method was the same as that used thruout the printing industry for this type of machine.

However, this task had been causing arm and wrist disorders among the printer operators for years, including burning and tingling sensations in their hands at the end of a shift. No one had reported an injury or even complained about the job, since "this is the way we've always done it." It was only when the facility ergonomics committee started to survey all jobs that the problem was uncovered.

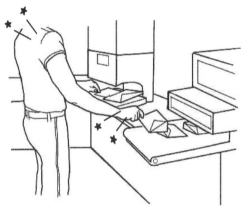 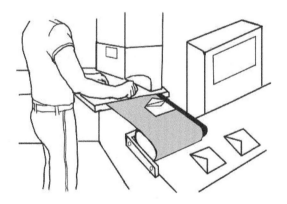

Before: Manual flipping   After: Mechanical flipper

Improvement — The solution was installing a semi-automatic flipper, adapted from a wholly automated machine engraver. The device eliminated about 5000 shoulder motions per day and was immediately popular with the employees.

The new device eliminates the repetitive work for both the left and the right hands. The operator was freed to concentrate on preparing for the next production run and checking quality from time to time (involving minimal repetitive hand activity).

Finding the solution

The idea came about when I asked about an automatic printing machine located nearby. The plant manager stated that "We use the automatic as often as we can. But we do a lot of short runs, and it isn't efficient to set it up for such a small volume." I looked closer at the automatic machine and noticed there was a mechanical flipper built in, since the cards and envelopes still needed to be flipped. After some discussion, we realized a similar mechanical flipper could be fabricated and attached to the manual presses.

This experience is not uncommon and provides a strategy for developing inexpensive improvements. Often, the expensive part of the automation is duplicating human decision-making. Duplicating the mechanical motions is much easier. So a good approach is to use the inexpensive mechanical devices that eliminate human motions, but retain the human for the decision-making. Use machines for what they do best and humans for what they do best.

This approach makes little sense if your goal is just to reduce headcount. But if your goal is to reduce injuries and workers' compensation costs while improving efficiency, it can be an excellent strategy.

Non-value-added

It is crucial to recognize that flipping the cards was a non-value-added step that amounted to an average of about 2/3 of the operation cycle time. It was a hindrance to output, since the operator had to stop productive work to stand and flip the cards and envelopes. With the mechanical flipper, the operator was freed to check quality without stopping production, plus had time to prepare for the next set up.

300% improvement

The cost of each mechanical feeder was insignificant, but it created dramatic differences:
- reduced the number of arm motions from 5000 to 0
- increased output from 5000 pieces per day to about 15,000 (300% increase)

Note that in some ways, this example is not very unusual. Most power tools like electric drills or powered lawn equipment also reduce motions while simultaneously increasing output.

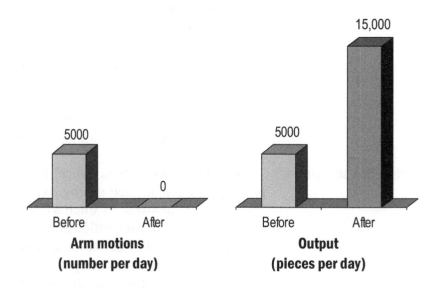

Arm motions	Output
(number per day)	(pieces per day)

Paper counting

Task — The task involved counting stacks of folio-sized paper and inserting a slipsheet every 50th or every 100th sheet. To accomplish this, the employee manually counted the paper in groups of 25, lifted the group of paper from one pallet to another, then placed a slipsheet on the counted groups as appropriate. The pallets were placed on the floor and a pallet jack was used to transport the product.

Ergonomic Issues — The task involved 45,000 to 50,000 finger motions per day to count the paper. Additionally, it required holding the non-dominant hand in a static pinch grip to support the paper while counting. Also, both arms had to be held in awkward, static postures while counting. Finally, the employee often had to work while bending over to reach the lower half of the stack of paper.

Employee morale was affected: "No one wanted to do this job."

Before: Manually counting stack upon stack of paper

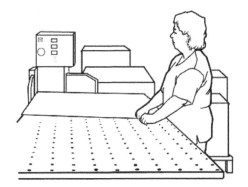

After: Workstation with lifts, paper counter, and "air table"

Improvements
- A semi-automatic paper counter was purchased. This machine was designed so that rough groups of paper could be slid into the counting machine, and then slid out in precisely-counted groups.
- A small, adjustable-height lift truck was purchased to hold up the incoming pallet of paper. This lift truck was used both to transport the pallet of paper and to raise the pallet to a comfortable height.
- An "air table" was installed to hold the groups of paper to be counted. The paper could now be slid, rather than lifted. An air table consists of a hollow top with pressurized air forced through small holes in the work surface, which thus reduces the friction of the paper pulled across the table surface. The air table was recycled from another operation that had been remodeled and thus was cost-free.

- The above equipment was placed next to an unused scissors-lift table that had previously been recessed into the floor. A special work area for counting was thus created.
- In order to reduce long reaches while counting the paper, the control switches for the lift truck were mounted on an extension cord.

These improvements reduced the finger repetitions to near zero. (A small percent of extra thick paper still had to be counted by hand, since it did not fit into the counting machine.)

The working postures of the lower back, arms and wrists were all improved. The employee was now able to work in good "neutral" postures for most of the day and was furthermore not restricted to working in a static position for periods of time. The changes also reduced the exertion needed to manipulate paper as well as eliminated the need to lift the counted groups of paper.

Employee initiated

A particular beauty of this case example is that it was the employee herself who developed the ideas. Her formal exposure to ergonomics was only slight — a one-hour training session held at the plant, plus reading and hearing about the ergonomics program in the company newsletter. She was aware that there was equipment available near her work area that was not being used.

Results

These improvements reduced finger motions from 45,000 per day to near zero. The counting time — all completely non-value-added — was reduced from 5 days per week to 2½ days, freeing up the employee to attend to other less arduous and more productive duties. The new equipment cost about $19,000 with a payback period of about one year.

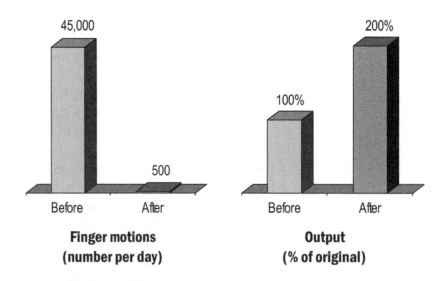

**Finger motions
(number per day)**

**Output
(% of original)**

Ribbon tying

About 20 employees had jobs that consisted of tying ribbons on baby birth announcement cards. The task was performed much as it would be done at home, sitting at an ordinary table and tying ribbons by hand. There were many fastidious hand motions and sustained pinch grips, and consequently, the number of wrist injuries in this area was high.

As with the previous example, one of the employees attended a short ergonomics class. That night, at home, she fabricated a prototype tying device using a manila folder, paper clips, and wrapping tape.

She brought her prototype to the plant the next day to show the rather astonished plant engineer (he had been unsuccessfully trying to develop a much more complicated machine for a few years). He fabricated more polished versions of her prototype and all the employees started using her device.

The inventor of the device was an ordinary employee of the sort that fill our workplaces and who can make marvelous contributions when systems are put in place to tap into their creativity.

Results

This unique device increased output 38% while cutting hand and wrist problems in half. Both were directly attributable to the fewer hand motions needed to tie the ribbons.

Before: Tying ribbons by hand

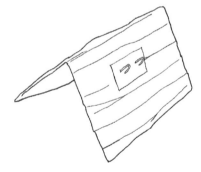

Prototype: Paper clips, a manila folder, and clear tape.

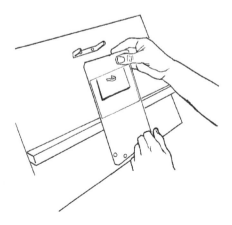

After: A more polished device

A block of wood

As a final example from this company, one of the facility ergonomics committees noticed an employee hunched over at a bench-mounted machine during one of their walkthru surveys. The bent neck and the bent wrists caught their eyes. The employee said that indeed her neck and wrists ached a bit, but nothing serious.

The committee members videotaped the job, then went to the conference room to watch the tape and brainstorm. After some discussion, someone suggested tilting the machine up at the back end. They returned to the job, tried tilting the machine, and got the employee's immediate feedback that it was better. Later, the maintenance department cut a wood block to serve as the support, plus added some stabilizers.

Tilting the machine simultaneously improved the employee's wrist and neck posture and made it easier for her to see what she was doing. The awkward posture may or may not have resulted in a recordable injury. But for the price of a block of wood, the task was improved.

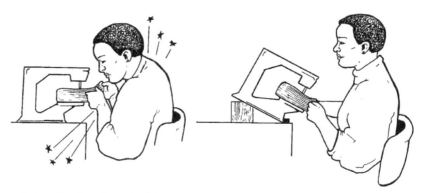

Before: bent neck and wrists   After: a block of wood to tilt the machine

Comments

The improvements to these jobs can seem simplistic and self-evident after the fact. However, the shortcomings of the tasks were either unknown or taken for granted by supervisors and engineers until the ergonomics process was brought down to the level of the employee.

None of the above jobs were cited by OSHA. Very likely none of these improvements would have been identified had it not been for soliciting employee ideas and systematically evaluating *all* tasks performed in the company's operations.

As one of the engineers from this 200-year-old company later commented to me, tongue in cheek, "You're right. Once every 40 to 50 years you should take a look at what you do."

On a personal note, this is one of the major reasons why I have been attracted to the field of workplace ergonomics — witnessing ordinary employees come to the fore with brilliant ideas. You just need to develop a workplace process and provide a little inspiration.

Discomfort surveys

This company administered annual discomfort surveys to all employees as part of their ergonomics program. In general, results showed positive changes in those areas where improvements were made, and no change in those areas where tasks had not yet been addressed at the time of the surveys.

The survey results proved useful for tracking accomplishments and for highlighting areas where additional work needed to be done. In this company, the surveys had an unanticipated effect of helping management keep a focus on employee concerns, since no one could deny that problems existed in their own work areas. Also, senior management appreciated the positive survey results, since it helped show that they were spending their money wisely.

Overall results

Injury reductions — The lost time MSD cases dropped dramatically and almost immediately. These cases are significant because they represent the most serious ones for employee well-being and because they are the most costly.

The "OSHA recordable" MSDs initially increased because of heightened employee awareness, then decreased. MSD "restricted workdays" also rose initially and then decreased. These trends appear to be the classic pattern of an active medical program.

Costs and benefits — In this company, careful track was kept of overall expenditures on ergonomics as well as the benefits. A good portion of the costs were investments in automated equipment (that is, more than just the low and medium cost improvements which have proved so effective elsewhere.) The benefits were mostly workers' comp savings, but included some productivity improvements as well.

After five years experience, total investment is estimated at $2.5 million and total benefits at $3.5 million. Note that, again, the costs are primarily one-time costs, while the benefits continue to accrue year after year.

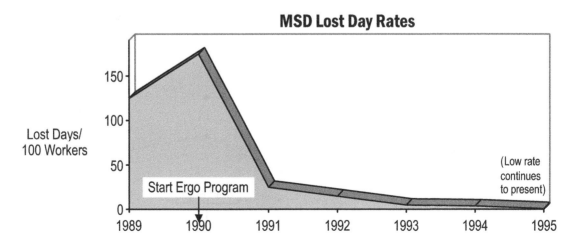

MSD Lost Day Rates

Lost Days/ 100 Workers

150

100

50

0

Start Ergo Program

(Low rate continues to present)

1989   1990   1991   1992   1993   1994   1995

Case Example 3
A Well-Prepared Manufacturing Division

This example is from one division of a major manufacturer. Perhaps the most noteworthy comment regarding this operation is that the success stemmed from a single week of training for facility teams. This organization knew what to do and needed no follow-up support whatsoever.

Background

The division manufactured metal components for the oil industry and employed roughly 4000 people in eight facilities. Teams of three or four people from each of the sites attended the initial training session where they were taught how to:

- Conduct task evaluations
- Focus on low-tech improvements
- Set up a common-sense process
- Provide training sessions for supervisors and employees

The primary motivation for instituting ergonomics programs was concern for employee well-being, but with an eye toward their rising workers' compensation costs. The plants were prepared by having previously instituted quality and process improvement programs and were ready for team-based ergonomics. The training fell on very fertile ground.

Results

The graph below shows workers' compensation cost data for the division in cents per hour per employee, plus equivalent data for the industry. The graph shows that the division had costs that were initially lower but rising faster than the industry average. After the ergonomics program, the division's costs dropped while the rest of the industry continued to rise.

These are powerful data, since the industry average serves as a type of control group and the cost per employee accounts for fluctuations in headcount.

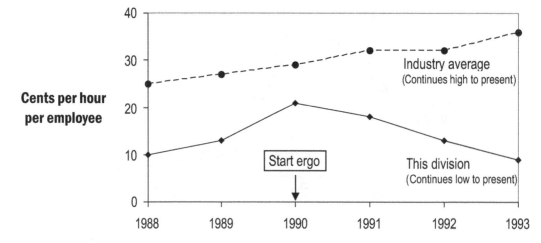

Workers' Compensation Costs

Results (continued)

Total workers' comp costs for the division are shown below. The company estimated that it saved about $12 million dollars in the first five years compared to the projected costs if they had done nothing. The investment costs for equipment purchases, maintenance modifications, and meeting time were not carefully tracked, but were certainly no more than $1 million, and likely much less.

Workers' Compensation Costs

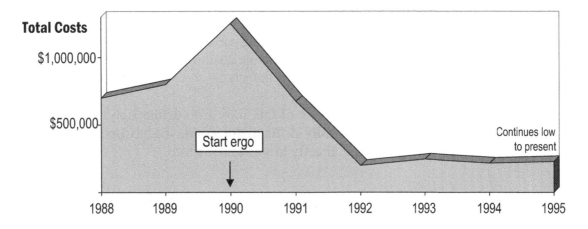

Task improvements

The improvements made in these plants were primarily low-tech and low cost. One very simple example as shown in the illustrations below is raising the height of a workbench to eliminate the need to work hunched over. Note that the "after" workstation is not perfect, specifically because of the bent neck. However, his neck posture is greatly improved compared to the "before" illustration, where his neck is parallel with the horizon. Moreover, his back posture is considerably improved.

No cost or benefits were quantified here because the improvement involved adding some extensions to the table legs costing less than $100. However, it is easy to imagine that the employee would be less fatigued and in a better position to do his job more effectively.

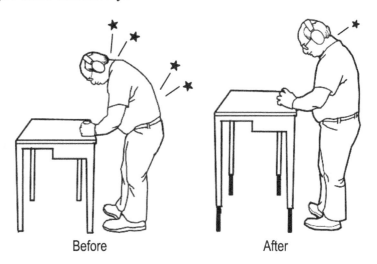

Before   After

**Quality improvement:
1500% ROI**

One of the especially useful examples from this company helps to demonstrate the positive impact that the ergonomics process has on reducing defects. One of the tasks involved standing at a dip tank and manually agitating the product in the liquid for several minutes. The initial concern that led engineers to the problem was the strain on the employee's arm from the repetitive shaking motions.

In response, they fabricated a small mechanical shaker to relieve the employee from this burdensome task. Subsequently, they discovered the number of defects in the product dropped considerably. The reason? It was simply too difficult and fatiguing for an employee to repetitively agitate the product sufficiently. A one-time cost of $400 for the shaker saved about $6000 per year in defects — a 1500% return on investment in one year!

The morale of this story is to deliberately seek out tasks that are physically difficult for employees to perform. The search can lead directly to sources of waste.

Costs and Benefits of Mechanical Shaker

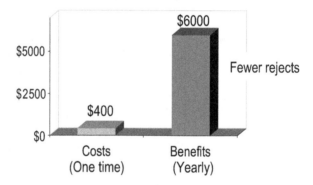

Comments

Communications — One key ingredient for the success within this manufacturing division was communications between sites. Operations were fairly similar and specific improvements could easily be shared.

Focus on low tech — Almost everything about this process was simple and low tech. One of the few formal measures was an audit of the sites to help insure that the committees were active and the problem-solving process in place.

Evaluation metrics — The primary measurement of the plant activities during the initial years was an audit of the workplace process performed by the division safety manager. In other words, as with the other companies described in these case examples, there was little or no emphasis on elaborate task scoring systems or other quantitative methods. The focus was on the process of plant teams evaluating jobs and brainstorming improvements.

Case Example 4
Meatpacking: How Ergonomics Changed an Entire Industry

It can be helpful for you to learn a bit about the experience in the meatpacking industry for several reasons:

- The industry has served as a testing ground for many aspects of the process of MSD prevention. For example, certain medical management practices were applied on a wide scale in the meat industry before elsewhere.

- Meatpacking came under government scrutiny prior to other industries, in particular with the publication of OSHA's *Ergonomics Program Management Guidelines for the Meatpacking Industry.* This experience provides some indication for the rest of private industry regarding what to expect in the future from OSHA.

- Health and safety experts and ergonomics coordinators may be interested from a purely professional point of view how this industry has been able to reduce these injuries.

The meatpacking industry has higher rates of hand and arm MSDs than any other.[*] From all indications, this statistic is true in all countries that have large meat industries, such as Denmark, Australia, and New Zealand. In the U.S., these high rates led to a series of OSHA citations in the late 1980s and the publication of the *Meatpacking Guidelines,* which prompted the companies to adopt major initiatives in ergonomics, which have paid off in many ways.

Injury Rates

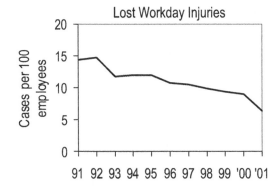

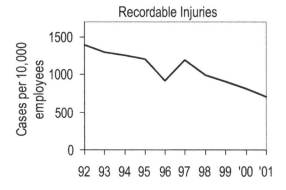

Source: U.S. Bureau of Labor Statistics 2005

---

[*] In 2002, when the U.S. Bureau of Labor Statistics (BLS) started to include low back pain, hernias, and other overexertion injuries in its calculations along with hand and arm "repeated trauma," the meat industry MSD rate dropped to almost average. See comments in Chapter 3, *Understanding MSDs.*

Results

Lower injuries — As shown in the graphs on the previous page, the injury rates for the entire meatpacking industry have dropped about 50% in the past decade.

Injury reductions in individual sites with especially good programs are even more dramatic, as shown below. Note that the rapid reduction in lost days in this case was attributed to a combination of a change in the workers' comp administration and improved medical management, as well as ergonomics.

Reduced turnover — One of the biggest effects of ergonomics in the slaughter industry has been a marked reduction in turnover. These are tough, physically demanding jobs involving tasks that many people would not want to perform under any circumstances — gutting hogs, cleaning beef intestines, etc. Improving the working conditions has made the meatpacking industry a more desirable place to work.

An example of the reduction in turnover is also shown below. These data are from the same facility as the injury data.

Individual Plant Results

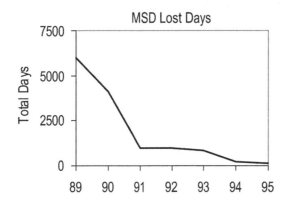

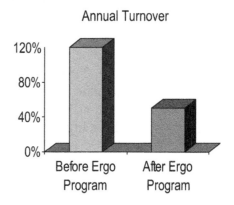

Why the meat industry?

Thruout history, the slaughter industry has always been hard and heavy work. Even today, this work is more physically demanding than most other industries. The reason is that the standard approaches to automation that are taken for granted in a car assembly plant, for example, cannot be used in meat plants, for two reasons:

1. Meat products come in odd-shaped and varying sizes, which make mechanization difficult. The meat industry stands in contrast to the auto industry where the exact shape, size, and location of most items are known with precision. Thus, the techniques of automation that were introduced into general manufacturing in the 1950s are still not possible in meat plants.

2. Sanitation requirements in the meat industry make any equipment development much more difficult. It is challenging to build equipment free of nooks, crannies, and porous materials that would be the breeding ground for bacteria.

Moreover, equipment in meat plants needs to be washed down regularly with high-pressure hoses. Most equipment used in general manufacturing would not last long in a meat plant. Electrical components would short out and metal surfaces would rust; in a meat plant, most equipment must be watertight or made of stainless steel.

Consequently, the design constraints for equipment in the meat industry are severe and opportunities for automation have been few. As a result, a higher percentage of the workforce performs manual work in the meat industry than in other industries, which is reflected in the MSD rates.

Task Improvements

Many personnel in the industry made statements like "We've made more changes with ergonomics in the past two years than we have in the past 20." And with those changes came lower costs of production. The following are some examples of creative solutions and concepts. Many of these improvements have implications beyond the meat industry.

Innovative ways to reduce force — A big problem in packing plants is the exertion it takes to use a knife to cut thru the meat. Creative solutions for this task include:

- For boning hams, companies have replaced a section of the conveyor for the hams with a tumbler (a rotating cylinder much like a cement mixer). As the hams tumble their way along, they become softer and are much easier to cut when they emerge.
- For beef, it was discovered that air or nitrogen injected between the muscle seams in a side of beef, makes the meat much easier to cut.
- Meat cooler designs have changed so that the meat tends to be more tempered and less frozen when cut.

The lesson here is that it wasn't the task of cutting meat per se that was changed. Rather, it was how the product was treated upstream in the process that made the jobs easier and less injurious. This is a point to consider in other operations.

Knife maintenance — Learning to keep a knife razor sharp is a crucial skill for a meat cutter. A dull knife increases exertion and thus raises the risk for an MSD of the hand and arm. But as anyone who has ever tried to learn how to sharpen a knife knows, this skill is not easy to learn.

Meat companies have had to place special emphasis on training employees in this skill, including: special trainers on the production lines, supervisors alert to new employees who may be having difficulties, spot checking of knives to test if they are sharp enough, and special work areas for new hires until they learn the skill. These steps provide guidance for all industries in how to train people in subtle work techniques.

Task improvements (continued)

Mechanization — Despite the constraints of sanitation and odd-shaped products, there were in fact many breakthroughs. One example is cleaning beef neck bones, which previously had been very labor intensive with highly repetitive wrist and arm motions. As a result of the ergonomics programs, the search for innovative methods led to a way to mechanize this work.

Overhead conveyors — A common problem in meat plants had been the height of overhead conveyors, traditionally placed quite high, thus causing unnecessary arm motions when tossing the meat or bones upward. Ways were found to lower the conveyors and even place some conveyors *under* the cutting tables so that upward motions were reduced.

Holes in tables — It is traditional in the meat industry to cut holes in work surfaces so that trimmed meat can easily be slid and taken away without the extra motions of picking up and putting down the meat. This is an example of a simple solution that can be applied in other industries.

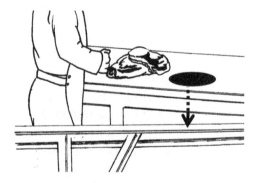

Adjustable-height work platforms — As a side note, perhaps the most unique work platforms in industry are the ones that have been used for decades in meat plants for splitting the sides of beef and pork. These platforms rise up and down about 10 feet while simultaneously moving sideways synchronized with the moving line. Platforms of this type could solve problems in other industries.

Ham boning line

Task — Hams are often boned on a moving line, with several employees making a few cuts each. It's exactly like a typical assembly line, except in this case it is a "disassembly" line.

Ergonomics issues — On a traditional boning line there are typically a number of issues:

- The cutting surfaces are horizontal, such as with a normal table or workbench. Consequently, employees need to hunch their necks and backs in order to access the hams.
- Heavy strain on the non-dominant hand to hold and manipulate the hams.
- Awkward wrist posture for the hand holding the knife.

Improvements — Some companies have started to use innovative designs:

- The cutting boards are tilted, which permits neck, back, and wrist posture to be closer to neutral positions.
- The product is supported by a simple fixture: a hook, which virtually eliminates repetitive motions in the non-dominant hand. This hook allows alternating between left and right hands or using both hands to hold the knife, thus reducing static load and exertion (as well as motions). Or for some cuts, fixturing the product enables use of a hand-held meat hook to pull muscle away from bone.

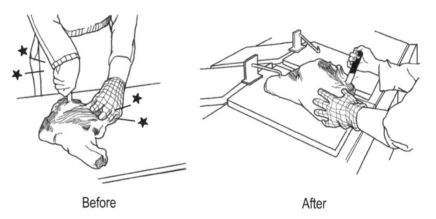

Before   After

Comments

The point here is not to make a comprehensive list on how to improve a meat plant. Rather the point is to provide a feel for the types of changes that have be made, as well as provide guidance for your own operations:

Optimism — If you can make improvements in as tough an industry as meatpacking, you can make them anywhere.

Cross-fertilization of ideas — Sometimes it is helpful to go outside your own industry to get new ideas.

Success — Trust the process, it works.

Trade association

As a final comment, the American Meat Institute (AMI) played an invaluable role in providing their member companies guidance and support. See page 160 for more information.

Case Example 5
Time Savings from a Pallet Lift

This example from a distribution center shows how a standard pallet lift reduced cycle time by 14–20%, plus reduced the load on the spine by 66%. The implications are universal and support the use of all types of pallet lifts and tilters — they eliminate wasted time and motions as well as reduce wear and tear on employees.

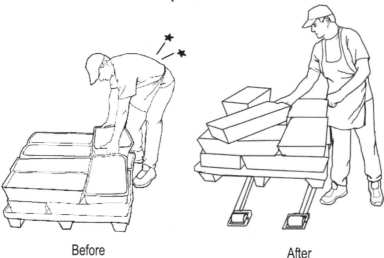

Before   After

Task

The situation in this case is very common thruout industry. The job involved lifting boxes from a pallet on the floor onto a conveyor belt. In this distribution center, the lifting was frequent and very repetitive.

The improvement was a pallet lift. In this case, the type of lift was portable, which allowed it to be used the same way as a manual pallet jack to move pallets. But when stationary, the lift could be raised up and down at the touch of a button, powered by a battery contained in the lift. Other than the portability, the lift was equivalent to other types of pallet lifts, so the benefits can be generalized to a variety of other types of equipment.

Evaluation method

The company asked me to quantify the costs and benefits. This evaluation technique is unique, but simple and powerful. The task was videotaped with and without using the pallet lift. As the completed tape was reviewed, it was paused each 0.5 second and the strain on the lower back calculated using ordinary formulas from the field of biomechanics.[*]

The results were graphed, so that time is displayed on the horizontal axis and the loads on lower back on the vertical axis.

---

[*] *Biomechanics* is the field of study used to estimate strain on joints. By knowing the position of the hands and any loads being manipulated, it is possible to calculate the effect on various joints and muscles.

**Understanding
the graph**

The graph shown below compares lifting a series of eight boxes onto the conveyor, first with the pallet on the floor and then with the pallet lift. The sequence and orientation of the boxes were exactly the same. The only difference is the height. The results are superimposed to help highlight the differences.

Each peak represents one lift. The lower the peak and the less area in the peak, the less strain on the back. The less horizontal distance at the base of each peak, the less time needed to make the lift.

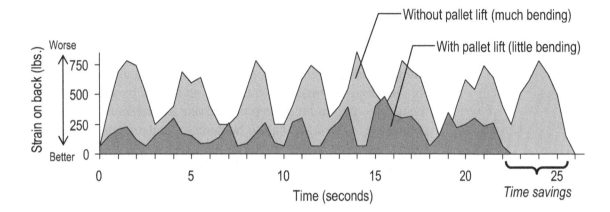

Results

The visual display of the graph clearly shows the benefits of the pallet lift and raising the boxes off the floor:
1. The peaks are lower, indicating less risk of injury
2. The area of the graph is less, also indicating less risk of injury
3. The time to perform the task is less, specifically eight trays are lifted using the pallet lift in the time it normally takes to lift seven.

Quantitative evaluation shows that the average load on the spine for these eight lifts without the pallet lift was 494.7 lbs. and with the pallet lift 166.1 lbs. Thus, the load on the spine was 66.1% less.

The time needed to complete these eight lifts was reduced from 25.5 seconds to 22.0 seconds, thus a savings of 13.7%. Additionally, a time study was performed on a full pallet-load of trays, which yielded a slightly larger time savings. The time needed to unload a full pallet of trays was about 6.5 minutes without the lift and about 5.2 minutes with the lift, thus a savings of about 20%.

Cost/benefit

The lift cost about $3000. The time savings for this high-volume lifting task was estimated at about $10,000 per year in this case, plus a very rough estimate of about $5000 per year in injuries. Cutting these estimates in half to be more conservative still yields a six-month payback for this facility.

Case Example 6
Favorite Low-Cost Solutions

These examples are from different operations, but all illustrate creative thinking. These examples reinforce the point that the workplace ergonomics process does not need to be difficult or expensive.

The typical problem is that we often take tasks for granted and do not question whether there is a better way, or indeed, if the tasks need to be performed at all. There is nothing magical about identifying and solving problems of this type. But having a formal process, with training and surveys, helps make it happen.

Maintenance crew

Task — In this chemical plant, the maintenance department had to remove a section of pipe periodically to access the inside for cleaning. At times it would take up to four people to perform the task — two to hold the pipe and one at each end to unbolt the connectors. The work had to be done in tandem, since if one end were uncoupled before the other, that end would likely drop and bind, thus creating considerably more work to unbind it. Even more difficult was lifting the pipe back up to reconnect it.

Solution — The crew attended a class in ergonomics and then was instructed to identify the "most miserable" job that they had to do and find a better way. They quickly concluded that they all hated to have to remove this section of pipe. After a few minutes of brainstorming, one of them suggested adding some wheels. End of story.

Results — The wheels were inexpensive and they made it possible for only one employee to perform the whole job, without ever needing to lift or manhandle the heavy pipe.

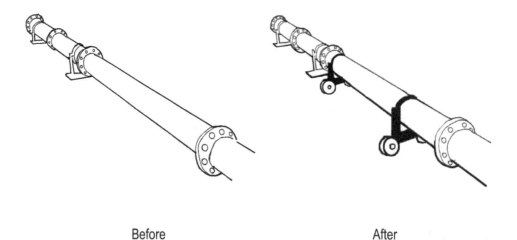

Before   After

A contraption

Task — The job was assembling high-end medical devices, which involved removing a small length of coating from a long string-like tube. The action was more or less like stripping electrical wires, and in fact, at one point the supervisor had tried a wire stripper, of the type that simultaneously strips off the insulation as it cuts. It worked, but the hand-squeezing motion was too strenuous to use for more than a few times. Consequently, the employees used a razor blade pinched between their fingers and laboriously scraped off the coating from the tube.

Solution — I went to a large hardware and building supplies store and walked the aisles while mulling over the problem. Eventually I spotted a small vise with a suction base that would attach to any work surface and could hold one end of the wire stripper. Securing one arm of the wire stripper would cut the exertion in half. I also remembered I had a short length of PVC pipe at home, which could be placed over the other arm of the wire stripper to serve as an extended lever.

Results — For a cost of less than $30, the contraption yielded an 80% reduction in hand motions, plus eliminated the pinch grips entirely. The result was a 60% reduction in the time it took to remove the coating.

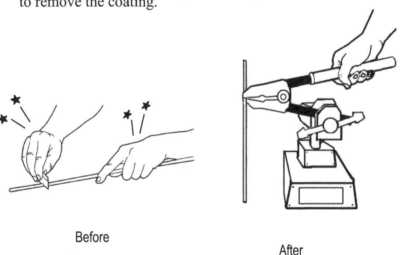

Before

After

$8000 vs. a block of wood

Task — The problem here involved loading and unloading metal castings on a cut-off saw. The layout required the operator to reach out and hold the heavy casting while tightening a crank to hold it in place. Bending over in this static posture with a heavy load caused a considerable amount of strain on the employee's lower back. The engineer was considering the idea of developing a hydraulic shuttle to facilitate loading, with an estimated cost of about $8000. He asked me to help justify this cost.

Solution — I suggested to the engineer and the operator that we brainstorm a bit more. After about 45 minutes and much discussion, we realized that a small block of wood would solve the problem. There was a two-by-four in the area, the employee had a wood saw in his car, and the improvement was implemented instantaneously.

Comment — The shuttle would have been better, but it is hard to justify $8000 when free works almost as well. The block of wood served as a *cradle*, an old-fashioned and low-tech way to support a load. It is important to always keep pressing for simple, low-tech fixes, even after you have identified something that works. You need to mull things over a while.

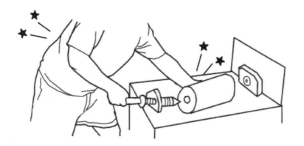

Before — Reaching and holding while simultaneously clamping

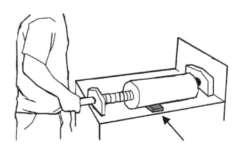

After — Supported by a block of wood

Postscript

In 2003, I was in Dubai, United Arab Emirates, making a presentation at a conference. The participants were medical and safety professionals from major oil, transportation, and refining companies. I asked why they were interested in ergonomics, since there were no government regulations, no expensive workers' compensation costs, and no labor unions. Their answer was "We are world class companies and this is what world class companies do."

Part II — How?

Chapter 5
Setting Up an Ergonomics Program

The intent of a program is to initiate a core set of actions in each workplace so that eventually ergonomics is fully a part of normal worklife.

Program Elements*

There are several basic elements of an ergonomics program that every employer should adopt in one form or another. This framework provides an approach for integrating ergonomics into day-to-day activities even among different sizes of employers and in different types of industries.

1. **Organization** — a plan for getting organized and involving people.

2. **Training** — an effort to provide training in ergonomics to people at all levels of the organization.

3. **Communication** — systems for communicating activities and progress.

4. **Task Evaluations** — a systematic way to review all work areas for needed improvements.

5. **Making Improvements** — the key part of the process; making improvements whenever feasible.

6. **Medical Management** — procedures to recognize and treat employees with symptoms of MSDs at as early a stage as possible.

7. **Monitoring Progress** — ways to measure and evaluate the program.

---

* These program elements represent my personal "standard" or perhaps "organizational checklist." Whenever I work with an organization, large or small, I go thru this list and, in conjunction with people from that organization, figure out what is appropriate for them. This simple approach is all that many successful employers have been using for years.

I have found that to try to write something that is more specific than this single page results in specifying activities that are not appropriate in some facilities. Incidentally, this framework was originally developed in 1987 and thus precedes the development of the various government documents that are outlined later in this book.

Options in Setting Up Your Program

There is no "one best way" to set up a program, as long as these basic program elements are implemented in an appropriate fashion. These elements are like a seven-legged "stool" that tips over if any one leg is removed.

The specific characteristics of an ergonomics program depend upon the needs and resources of each facility. Individual companies vary in size, complexity of operations, and system of management. The level of activity and resources committed to the program depend upon the extent of existing problems. In small organizations, these program elements can be achieved with relatively modest efforts, and indeed, a trade association can do much of the work. In large organizations, or ones with severe problems, considerably more work may be necessary. The intent of these materials is to raise the key issues in a way that can be modified in a feasible way to fit the unique circumstances of every work site.

Program Evolution

Programs can and should evolve. The kinds of activities that are often needed during the initial stages may well be different from those that are appropriate later, as sophistication matures. Part of this evolution is inherent in the growth of any organizational activity and part of it is based on the simple fact that the needs of any particular employer can change with time.

Tips and Ideas

The following sections provide tips, ideas, and further background for setting up a workplace process based on this framework. These are essentially best practices from a large number of different workplaces in many industries. This kit also contains forms and worksheets that you can use to help implement these practices.

Everything is based on years of practical experience in industry, including developing large-scale programs in some of the largest employers in the U.S. and in industry-wide activities, including the meat and poultry industries.

This material also serves as a supplement to the various standards and regulatory activities that have emerged. Note that it is not the intent of this kit to repeat or interpret these guidelines and regulations. Rather, the intent here is to highlight and expand on certain issues and provide additional ideas and guidance for setting up a program.

Note that preventing work-related MSDs is the focus of this material, but the same process can be used to improve other workplace efficiencies.

Organization

Ultimately, ergonomics should be such a part of normal day-to-day activities that there should be no need for any type of special organizational planning. Employees, supervisors, architects, engineers, and designers should eventually have the question "how does the human fit in?" so ingrained in their thinking, that consideration for ergonomics will be automatic.

However, at this point in the evolution of workplace ergonomics, most employers probably need to take special actions to advance the process. Typically, adopting new tools and concepts into your system doesn't just happen.

The first step is to define an overall organizational structure. Ideally, a plan (or, at least the first phase of an overall plan) should be in place before any other activities are started.

Management responsibility

Define authority — Assigning responsibility is typically one of the first actions that needs to be taken. Someone needs to be in charge and everyone needs to know their role. Authority must be defined so that people active in the ergonomics program know which decisions they can make and which ones must be brought to senior management. Some way must be found to make sure that personnel assigned to certain tasks have the authority to do so and, conversely, a way to follow up to make sure they have performed their duties.

Find a "champion" — The ultimate responsibility rests with line management, but usually the project activities are delegated. Yet, even when the responsibility is delegated, someone from senior management needs to stay involved to some degree. It is common to identify a senior management "champion" or "sponsor" to promote the ergonomics effort. A "champion" can mean different things in different companies, but the idea here is to have someone high up in the organization represent the interests of the ergo project.

This type of champion is often not involved at all in your day-to-day activities, so the ergonomics coordinator must keep this person briefed and aware of key issues. There may need to be a regular meeting time between the coordinator and the champion to review issues, especially prior to larger meetings.

Commit resources to address the issues. Examples of such resources include:

- assign staff time (for example, 50% of someone's time)
- provide an ergonomics budget
- retain an ergonomics consultant
- assign maintenance personnel to ergonomic fixes
- provide for employees to take time off from their jobs to be involved in the program

Commitment

Be visible. Actions speak louder than words. Personnel must be able to *see* that top management is committed.

(In one well-known company that I'm familiar with, the senior corporate executives spend a week each year working in one of their facilities — and the union president in that facility gets to pick the jobs!)

Demonstrate support in other ways. "Commitment" can be difficult to define or measure in any one particular activity, but is often very evident in its overall effect. In part, management commitment means to:

- support activities
- provide resources
- be knowledgeable and up-to-date
- be timely in response to requests and recommendations

Accountability

Establish systems to ensure follow-thru and continuity of your actions. How accountability is created and enforced can vary considerably in different companies, but one way is to develop some type of management-by-objective (MBO) system. You do this in the following way:

- Think thru what you want each affected individual to do and write it down. Some of this will be generic, but much of it can be very specific, for example, what you want the supervisors in Departments 14 and 21 and their superintendent to accomplish in the next three months.
- Provide this list to the senior manager. Usually the ergonomics coordinator would do this either together with the champion, or arrange to have the champion meet with the senior manager independently.
- The senior manager then discusses the objectives with the superintendent, who in turn discusses the objectives with the supervisors. Three months later, the senior manager reviews with the superintendent what has been done.

Find what works for you. The measures necessary to ensure accountability can be formal, like the above system, or very informal, depending upon the normal management style of the organization and the circumstances at hand:

- In some companies, the objective is for the supervisors to implement one improvement per quarter.
- A more sophisticated version is to have the supervisors reduce, for example, risk ratings for high priority jobs.
- Sometimes the senior manager needs to get involved only rarely, such as when some element is resisting change.

Keep in mind, too, that in many organizations, accountability is not a problem at all. The guidance above is in case an issue ever arises at your site.

Workers' comp chargeback

An effective way to build accountability is to charge back workers' compensation costs to each department. Usually the costs are sufficiently high to provide incentive and the chargeback places these costs into the regular system of accountability.

The chargeback is important because workers' compensation is typically considered to be corporate overhead, something that the Human Resources Department pays and not a particular concern to supervisors and other department managers. By charging back these costs to the departments where the losses were incurred, it helps place accountability where it belongs.

This system can vividly change the financial calculations that managers and supervisors make. Instead of trying to decide whether to buy a $3000 pallet lift or not, the decision changes to whether they should buy a $3000 pallet lift or take the chance of a $50,000 back injury.

First line supervisors

Whether there is a chargeback or not, it is important that supervisors are involved and know their responsibilities. More details are provided in Chapter 11 *Training Slides and Script*, but a good summary of responsibilities is that managers and supervisors should:

1. Know the basic principles of ergonomics.
2. Understand the role of ergonomics in identifying wasted motions, unnecessary activities, fatiguing conditions, and other sources of inefficiency.
3. Understand the basic problem-solving process.
4. Know about musculoskeletal disorders — general types, symptoms, and causes.
5. Recognize the ergonomics issues in your area and activities that have caused musculoskeletal disorders.
6. Work with employees to make those changes that you can do by yourselves. Communicate with TeamErgo regarding improvements.
7. If an employee reports symptoms, or you see an employee who appears to have symptoms, make sure the employee receives medical attention.
8. Show concern for employees and their well-being. Talk with employees about ergonomics and solicit input and requests.
9. If an employee has work restrictions, help accommodate the employee and see that medical restrictions are observed.
10. Be familiar with company policies and programs for ergonomics and musculoskeletal disorders.

Written plan

Reducing your program to writing helps clarify goals and activities, as is routine in any business operation. Moreover, the plan serves as a tool for communication. Many safety professionals consider this step the cornerstone of an effective safety and health program.

More details and a worksheet are provided later in this book, but essentially, to write this plan:

- Review each item of this section.
- Determine what is applicable to your operations.
- Define in writing the specific approach you are planning to take in addressing these items.
- Identify target dates for:
  - implementing the various elements of the program.
  - making specific engineering improvements as your program progresses.
- Update the plan periodically to adapt to the results of your experience, to changing circumstances, or to identify future objectives.

Ergonomics team

A team approach generally works best for ergonomics, rather than just assigning the responsibility to one person.[*] Often it involves forming a new group — "TeamErgo" — but it could be an existing one.

Typically, the people involved should represent the various site functions needed to coordinate and implement changes, as well as to provide different perspectives on issues:

- safety and medical
- human resources
- purchasing
- engineering, facilities, and maintenance
- operations management
- union or employee representatives

Employees AND management — Note that in most cases, management-only committees or employee-only committees are not very effective. Management needs to be involved to be able to make decisions. Employees need to be involved to provide input on issues and gain acceptance for change.

Coordinator

Someone needs to coordinate the various aspects of the ergonomics project. The coordinator can be chosen from any job function and there is no standard practice within industry for what kind of person is selected. Sometimes the facilities or maintenance manager is coordinator, especially when there is a large of amount of physical changes to equipment needed. Other times the nurse can be effective. On occasion, the site manager wishes to take the role, while in other locations the duties fall quite naturally to the human resources manager.

Hourly employee — It is not unusual for an hourly committee member to serve as coordinator of the whole facility process.

---

[*] There are exceptions. For example, in smaller organizations, there may be no need for a team and one person can coordinate all activities.

This approach usually provides the great advantage of someone who is familiar with what actually happens on the workplace floor and who has credibility with other employees. Furthermore, in some situations, a primary function of the coordinator is to continuously check with various people such as engineers, purchasers, or supervisors on the status of various projects. This can be time consuming and there is usually no reason why an hourly person can't perform this function.

This approach requires good support and acceptance from management members of the team. If anyone in management is threatened by the very idea of an employee taking a leadership position, it might not work at all.

Team activities

Functions of TeamErgo typically center on the following types of actions:
- Set priorities for evaluating and improving jobs.
- Conduct surveys and evaluations, or assist individual supervisors and employees in specific work areas in doing so.
- Identify action items and who is responsible for follow-up.
- Coordinate activities.
- Establish procedures.
- Document and communicate activities.

Work between meetings — One effective practice in many organizations is to do most of the actual work in between meetings and then use the meeting time to communicate and plan. In other words, an individual or subcommittee might be given an assignment, such as finding a vendor for a lifting aid or developing a draft plan for a new layout of a work area. Then, this work is conducted before the next meeting is held. Thus the meeting time isn't spent doing time-consuming activities, rather to update everyone on results.

Sub-teams and temporary members

It can be helpful to divide TeamErgo into sub-teams so that individual members can focus on specific topics. Examples are communication, training, and concentrating on different work areas.

Simultaneously, it can also be helpful to add temporary members to the team. The primary instance is when the team is focusing on a particular work area, when it would be advantageous to bring in more people from that area.

Some companies have a small core group (six or seven permanent members) and a large number of temporary members. In this case, the core group becomes a policy-deciding body.

Central team vs. departmental teams

One common question that many companies face is whether to have a single site-wide team or multiple teams operating in various departments or on multiple shifts. The answer to this question obviously depends on the size and complexity of operations and the organizational culture of the employer. Nonetheless, there are still pros and cons.

Having only a single team is much easier to coordinate, but in some companies there is no alternative other than multiple teams. In the latter case, it is crucial to have good coordination, which can place considerable stress on the coordinator(s) in attending meetings on different shifts.

It helps to keep in mind that, ultimately, ergonomics should be a normal part of day-to-day activities in the work areas themselves. This objective speaks to the idea of promoting departmental teams. Perhaps a good perspective is to think of a stronger central team in the beginning stages of the process, with increasingly more emphasis on departmental teams as the system matures. Eventually, there might not be a need for a central team at all, since all the responsibilities would be handled by line management as part of their normal functions.

New or existing team?

In some facilities, existing groups, such as the safety committee or some type of process improvement committee, already involve the right people and it would be natural to add ergonomics to their agendas. In other cases, the top manager's staff supplemented with a few other key individuals offers the best approach.

Again, the point here is to think about what various options there are at your site — it makes no sense to reinvent the wheel and develop a new structure when there are existing organizational mechanisms that can serve the purpose.

One rule of thumb is to use whatever system has been working in the past. For example, if a safety committee has been effective, there is no particular need to change anything, other than add ergonomics issues to the list of concerns. If the safety committee has *not* been effective, then there is a need to think about alternatives.

Furthermore, if there are active "action teams" or "kaizen events" or whatever mechanisms, they can be used instead of a safety or ergonomics committee. There may not be any need to start up a new committee or new process.

This rule of thumb of using whatever has worked in the past can also help determine the size and composition of TeamErgo. What has been the best setup in your experience?

Team vs. line management

The most difficult part of planning is often determining the relationship between TeamErgo and line management. In some instances, line management can resent and oppose any attempt to "intrude" in their areas. In other situations, managers may do the opposite and abdicate their responsibilities, forcing TeamErgo to do their work. A balance must be sought, which is not always easy to accomplish.

The nature of these relationships also depends on the

industry. In light and medium-sized manufacturing, supervisors at the level of individual departments are central to any effort to change workstations. On the other hand, in other industries, such as oil refineries and chemical processing plants, many issues are beyond the control of any supervisor or group of employees.

It helps to think of TeamErgo as a resource to the departments, with accountability residing with line managers and the people in the work area themselves. In the end, the responsibility for ergonomics lies in the departments and line managers themselves, *not* a site-wide committee. Accordingly, the process shouldn't be driven by the safety department or TeamErgo.

A good approach is to consciously structure your program so that the team starts out strong, but eventually fades away as ergonomics becomes a part of everyday worklife.

Employee involvement

There are numerous advantages for involving employees in building an effective program. The people who do the actual work often have special insights into ways of improving their own jobs, especially if given training in ergonomic principles. Participation often helps pave the way for accepting improved methods and equipment. And, in many ways, it is simply common courtesy to consult with people if you are going to change their work areas.

Ways to obtain employee input include the following options:

- Set up a suggestion system.
- Interview individual employees at their workstations.
- Conduct an employee survey.
- Form department-level ErgoTeams, which generally makes it easier to involve more employees.
- Appoint employee or union representatives to the site-wide TeamErgo.
- Hold small group discussions when certain jobs or areas are being addressed.

Caveats

Remember, too, that employees aren't always right, even tho their involvement is crucial. We want to solicit ideas, but not every idea is a good one.

Furthermore, task design isn't exactly a democracy and at some point you may need to make judgments about best methods that may differ from what employees think. If you are convinced a particular technique is a better way, you may need to require people to work that better way. But be careful. There are gray areas here, where science and engineering cannot yet provide definitive guidance on what is the better way.

It can be helpful to distinguish between "personal comfort" issues and "best practices." The classic example of a personal comfort issue is air temperature. Some people like it colder and others warmer. Within reason, there is no correct temperature and it is thus very difficult, if not impossible, to resolve within a group of people who differ.

At the other extreme is the case where someone prefers to work in an unsafe way. If one method is clearly better and safer than another, it becomes a rule, just like any other safety practice. In fact, at some point, using the better work method can and should become a condition of employment, with all the disciplinary procedures that this involves.

Self-managed workgroups

Many employers in the U.S. are moving in the direction of self-managed workgroups, whether consciously or not. In this situation, responsibility lies within in the group, but the group often needs outside assistance with various technical issues, including safety and health.

No one can be expected to know everything. Once again, this speaks to thinking of TeamErgo as a *resource* with the ultimate goal of empowering the workgroups to make the right decisions.

Other ongoing initiatives

Think about what else is happening in your facility. There may be a push for improving the flow of work (i.e. "lean manufacturing") or there could be continued emphasis on quality improvement. You should make sure that these activities in ergonomics are integrated with other initiatives. Careful planning of this sort helps in two ways: (1) ergonomics does not appear as a separate or disjointed effort to workplace personnel, and (2) the ergonomics program takes advantage of and contributes to momentum generated by these other efforts.

Be creative. As with most everything else in this book, there may be no single best approach that works for all employers. The point is to reflect on different methods and choose that which works best for you. It might take some trial and error to develop the best practice for you.

A practical tool for HR

Ergonomics is a human resource manager's dream: a people-oriented process that can improve employee morale and cut injuries and workers' compensation costs. The process does not need to be technical and can easily be coordinated by an HR manager.

Much of the work in running an ergonomics program is related to facilitating discussions, an activity for which many HR personnel are well suited. HR people can also serve as the fresh set of eyes to ask the penetrating questions: "Why do you do it this way? Why can't you do it another way?" There have been many examples where HR personnel who are not familiar with production processes have played crucial and successful roles.

Training

Personnel at all levels in the organization should receive training in ergonomics. Everyone needs to know the rules of work.

Moreover, training in basic principles is necessary in order to enable people to provide their ideas and input in a systematic way. The extent and focus of this training may vary depending upon the audience, but even then there are certain common denominators.

Since most people are unfamiliar with ergonomics and MSDs, the thrust of much of this initial training is to introduce both managers and employees to the basic concepts and principles. The other component is to provide everyone with sufficient background so that they can evaluate and improve the use of their own workstations.

This section provides an overview of the training that needs to be considered. Furthermore, later chapters provide lesson plans, handouts, and visual aids to help you conduct training. Electronic materials, including PowerPoint presentations and short handouts, also accompany the lesson plans.

Types of training

MSD right-to-know — OSHA tends to think about training as a "right-to-know" issue, similar to hazardous communication regulations. That is, people exposed to MSD risk factors have a right to know that they are so exposed, what the symptoms of this exposure are, what their recourses are, and what the employer is doing to abate hazards.

The user-friendly workplace — Since ergonomics encompasses more than preventing MSDs, the following material can also be used in a broader sense — to encourage employees and supervisors alike to be innovative and come up with ideas for making their work areas more user-friendly. Thus, workplaces without significant MSD problems can still benefit from the training.

Procedures and programs — People need to be informed about your employer's various procedures, for example, what to do when you want to report a problem. Also, it is important to let people know what activities and actions management is taking on various topics, including how you are addressing the ergonomics issues in your facility.

Job skills — Employers need to make sure that everyone has the basic understanding of their jobs to do their tasks properly, using the easiest methods. Simply having supervisors work with each individual can go a long way in this area.

A more effective, new technique is to videotape the most highly skilled employees (or every employee), then hold small group sessions for the people to view themselves at work. This activity can set the stage for the employees to help each other determine the easiest ways of accomplishing a task.

Tool and equipment maintenance skills — A closely related issue is ensuring that people know how to maintain their tools and equipment. This issue is important since poorly maintained tools often make it harder for employees to do their jobs (for example, keeping knives sharp in the meatpacking industry). Thus, specialized training is sometimes needed, both for production employees as well as maintenance personnel.

Personal injury prevention strategies — Finally, additional training about personal responsibilities and habits can be important, including:

- body mechanics and good lifting practices
- wellness and exercise

General agenda

Everyone needs to know certain basic concepts, regardless if they are line management, support staff, or general employees. The general agenda for awareness sessions for either managers or employees is the same:

- What is ergonomics?
- What are musculoskeletal disorders?
- Ergonomic issues in your work area
- This facility's ergonomics program
  - TeamErgo: structure and members
  - Activities for evaluating and improving jobs
  - Medical issues and procedures for getting help
  - What you should do
- Option: Employee survey
- Option: Ergonomics and process improvement

Audience-specific training

In addition to these core concepts and principles, there is a need to provide more information on specific issues to various audiences. In some cases, there are also differences in the level of detail provided and the framework for addressing personal responsibility.

Top management needs to be briefed on the business benefits of ergonomics and on background items ranging from underlying strategic planning (aging workforce, rising medical costs, etc.) to current approaches to medical management.

An item to address in particular is the potential financial benefits from the ergonomics program:

- reduced workers' compensation costs
- lower turnover and absenteeism
- improved morale
- innovations
- improvements in quality and efficiency
- regulatory and litigation issues

Supervisors and other managers are an obvious good source of ideas for job improvements. A training session is an ideal time to solicit these ideas.

Supervisors should also be able to recognize the ergonomic issues in their area, especially those that may contribute to cumulative trauma cases.

And finally, supervisors should be taught to recognize when employees are having problems and what actions need to be taken. The role of supervisors in proactive medical management of MSDs is growing increasingly important.

Engineers, maintenance, and purchasing personnel have a key role in specifying, building and modifying equipment. They need to be sensitized to the discomfort that employees can experience if inappropriate decisions are made. Again, this group is a good source of ideas and should be encouraged to be innovative. Engineers may also benefit from advanced training in the analytic tools and quantitative methods of ergonomics.

Vendors who are involved in designing or installing equipment could also be invited to training sessions. Some vendors could give presentations on the use of their products and related background information.

TeamErgo members (or the individuals with primary responsibility for the program) need the most detailed training. This includes principles of ergonomics, how to administer the program, and developments within the field.

Specialists such as nurses and other health care providers need ongoing specialized training. An increasing number of professional seminars on musculoskeletal disorders are becoming available.

Employees need to have good awareness about their own roles and responsibilities in the overall process, how to set up their own work areas, and how to identify problems and ideas for improvements.

Job skills

Employees need to have a considerable amount of information that is specific to their work areas, including knowledge of issues that reduce risk of injury or inefficient use of their time.

Job know-how —Traditionally, supervisors have held the responsibility of providing training. Furthermore, the instruction typically consists of the basic steps and the requirements for completing the task and many of the details are left for the employees to figure out for themselves.

To help provide as much know-how as possible, many companies now use special trainers for this purpose, in addition to installing special training lines and equipment to enable better training. Instruction has also been modified to include much more information that helps from an ergonomics perspective.

Adjustability — An important example is making sure everyone knows what equipment is adjustable and how to adjust it. Similarly, people need to understand how to best lay out their work areas, even to the extent of being told that it is okay to move things if it makes the task easier.

Best methods — Another important consideration is the need for improved training to help employees use smooth and easy work methods. It is not uncommon to see one employee perform a task smoothly with minimal effort while the next employee struggles to do the same job. Thru experience, the first employee has gained skill and learned techniques that make the work easier and more productive. Consequently, new approaches to training may be needed, including these steps:

- Evaluate jobs to find the best work method. Analyzing the differences in work methods between individual employees or between different shifts or plants can provide valuable insight.
- Videotape employees at work and hold small group sessions for employees to view themselves at work. You can use these sessions to set the stage for employees to discuss these issues and share techniques.
- And finally, make sure you allow adequate time for employee training before the employee is expected to perform at full capacity.

Information on MSDs

Instruct employees in affected jobs on MSDs, the symptoms, and how they should report any problems. (In many ways, this training is equivalent to hazard communications training for exposure to chemicals, but for MSDs.)

The sooner employees report symptoms, the better the chances for effective (and less costly) treatment. Follow-up training may be advisable, such as annual meetings to review basic issues and recent developments.

Provide production employees with basic information on MSD risk factors and principles of prevention. These training sessions are a good time to solicit employee ideas and have a discussion of specific needs and possible improvements.

Tool use and equipment maintenance

Evaluate your tool programs. Sometimes an improperly used tool can contribute to employee problems. Follow-up with employees periodically to ensure that the skills have been learned. Be systematic. In the past, training in tool and equipment use has tended to be haphazard, at least in some organizations.

Review training aspects of equipment maintenance. Primary targets for training are personnel who maintain the equipment. Users may also benefit from training. Prompt repair of tools can reduce the potential for problems.

Contact suppliers. Many suppliers have developed good training materials for the proper use and repair of their equipment and are willing to provide in-plant training sessions.

Sequence and timing of training

It is not necessary or even beneficial to train everyone all at once. Usually there is a sequence to the training.

Supervisors: early — Typically, you should train supervisors at an early stage. By explaining benefits, goals and plans, this helps achieve a buy-in from a group of people crucial to a facility-wide effort.

Employees: later — Employee training should generally be done in the later stages of setting up a program, after procedures and systems have been implemented and some experience is gained with making ergonomics improvements. This helps avoid raising expectations for job improvements too high before the program can handle all the ideas that are typically generated as a result of the training.

Pilot projects — An exception to the above rule is if a pilot work area is selected for ergonomics task analysis and improvement. In this case, initial training ought to be provided to employees in this area, to explain to them the purpose of the pilot project and to gain their input on ergonomic issues and possible ideas for improvement.

Initial awareness — Another exception to this general rule is providing short communication sessions with all employees in the early stages of a program, simply to let them know that activity is beginning and that all employees will eventually be trained.

Follow-up sessions — The sessions described above concern the initial introduction of ergonomics to the workforce. Subsequently, there is a need for updates. In some sites where ergo issues are especially important, it can be valuable to hold special meetings for these updates. More often, the updates can be integrated into other existing programs, such as monthly safety talks.

A good way to provide these updates is to describe recent improvements in the workplace with a focus on the before-and-after changes. That way, people are kept up-to-date on events in the workplace, while reminded of key concepts and principles of ergonomics, all in a positive atmosphere of "here are things that have improved."

Presentation options

There are several choices of media with which to present the training. Often a combination of approaches works best, allowing for flexibility by using the best of several systems.

A PowerPoint presentation is often the tool of choice for skilled instructors. Slides are easy to customize — especially with digital photographs from your site — and the large, clear images provide the instructor with opportunity to point out and amplify details. The instructor can easily interact with the audience during the presentation. This is the least expensive approach and very easily customized. However, this format requires preparation and knowledge by the instructor.

Videotapes are easy to use, especially for presenting the core information of a program. Obviously, video-based training programs have become quite popular in recent years. However, they are expensive to customize for the particular issues in your workplace and lack the give and take of live interaction.

Self-paced computer training has the advantage of being flexible for everyone's different schedules. This method is useful for providing basic information, but usually cannot address the specifics of individual workstations.

Outside experts can bring credibility and a fresh perspective, plus in-depth familiarity with the topic and sometimes a polished "show." Furthermore, an ergonomist who is used for training can also help in reviewing workstations and developing other aspects of a program. The disadvantages are the expense, the lack in empowering your own in-house instructing capabilities, and the fact that not every professional ergonomist is an engaging trainer.

Monthly safety meetings

Most companies have monthly safety meetings or some equivalent that provides a good opportunity and an existing mechanism for ergonomics training and communication. You can use these sessions in a couple of ways:

Introduction to ergonomics — It is possible to provide the basic introductory material in a series of short sessions, rather than a single long one. To be sure, an introduction to a major topic like ergonomics is usually done in a single session, but breaking it up is an option.

Updates — An excellent use of the monthly safety meetings is for updates. Ergonomics will be part of workplace safety efforts from now on and there is plenty of information that you can provide in short safety meetings. A good format for presenting the material is success stories; you can show a video or a slide of some workstation before and after changes were made. This format simultaneously (1) educates people about what to look for and how to make improvements, and (2) communicates important news about things happening in their own workplace.

Problem-solving sessions

It is possible to take advantage of the introductory training session to discuss issues in the participants' own work areas and to problem-solve during the class. Discussions of this sort help in two crucial ways: (1) they serve to reinforce the training presentation so that people remember more, and (2) you can maximize the use of the time and actually fix some problems.

Taking this approach can expand the session to about two hours in length, since you would usually want to show additional videos or slides of your operations to promote the deliberations. Sample discussion questions are included in Chapter 11 *Training Slides and Script*.

Communications

Companies that have been successful in ergonomics have placed strong emphasis on good communications. Or to say the same thing in negative terms, poor communications is a frequent reason for the failure of many ergonomics programs. The reason that programs fail is often not, as is often assumed, the lack of management commitment or the lack of a budget to make changes. Usually, there is enough money and commitment to do *something*. The problem often is that the program is progressing well enough, but no one knows about it. After a while, people think that nothing is happening and they lose interest. And the program slowly dissolves.

Thus, good communication is essential for the success of a program. You must take time to sit down and develop your plan for communications. It's not hard and it shouldn't take much time, but it won't happen by itself. You need to toot your own horn.

With employees

Inform employees at the start of the overall objectives of the ergonomics program. This doesn't necessarily need to be the whole site, at least at first. But for pilot projects and initial activities, make sure you introduce yourself properly to the people. Explain why videotapes are being taken of jobs and why questions are being asked about aches and pains. Be sure to notify affected employees in advance when certain areas or pieces of equipment are going to be modified (that's just being polite).

Groom expectations by choosing your words carefully. Often you want to stimulate interest, but you don't want to raise expectations too high. Typically, you have leeway to say exactly what needs to be said.

Provide feedback to personnel who have provided ideas of the status of those suggestions. Even if the answer is "no," you still need to tell them. Most people would rather hear that their idea was turned down than not hear anything at all.

Provide updates regularly. This is the most important item. If people aren't told what's going on, they may assume nothing is. In particular, changes made in one part of a facility may not be known in other parts. Special steps are needed to keep everyone informed.

Formats

TeamErgo should list all the different avenues for communications that are available in the facility, then develop a plan for using those mechanisms. Typical methods are:
- **Bulletin boards**, with photos and testimonials by employees whose jobs have been improved. For example, you provide before-and-after photos of the job and get a quote from a

happier employee: "I used to have to bend over all the time at my job and my back was killing me, but TeamErgo raised my machine and now I'm better."

- **Newsletters** can be used in the same way. If every issue of your site newsletter contained just one success story, it would typically have a big impact. It's a positive event and readers learn a bit more each time.
- **Monthly meetings**. Most organizations have some type of regular meetings. TeamErgo members can prepare talking points or other materials for these meetings, or even make the presentations.

There are many other avenues that you can use: posters, notes in paychecks, notices in the bathrooms, health fairs, etc. Just make sure to have an actual *plan*. For example, include in your written plan that you will place one success story per month in your site's weekly newsletter. (Just saying that "we will use the newsletter" is not much of a plan. Saying that "John and Bill will write one success story per month for the newsletter" is a plan.)

Among decision-makers

You need to identify ways to keep all decision-makers aware of plans — managers, facilities staff, safety, purchasing, etc. Timely communication is essential to plan properly, coordinate activities, and build momentum. In addition to the items listed above for general employees that reach most everyone, you need to always keep the key players in the loop. You can arrange to have an agenda item in staff meetings, or you can make effective use of email.

This type of communication constitutes a huge part of the coordinator's job. You may need to touch base daily with a lot of people to learn the status of activities and to get their input.

Between company facilities

If you have more than one facility, make sure to share ideas and progress. The use of email, digital photographs, and digital video make this sharing very easy. You can also develop an intranet website where you can post your success stories. Finally, don't forget the old-fashioned ways like the telephone, site visits, and conferences. The prime payoff for these efforts is good cross-fertilization of ideas, which is key for innovation. The more alike your facilities are, the more important this sharing is.

The Minister of Propaganda

One way to do help insure that good communications happens is to assign one team member to be "Minister of Propaganda," a role designed strictly to focus on keeping the whole site up to date on developments.

Task Analysis

A good ergonomics process requires a systematic way to identify all ergonomics issues. Ultimately, your goal should be to evaluate every single task performed at your site. The following section briefly describes various techniques that can be used.

There are three levels of task analysis:

A. Where to start — background surveys
B. Task analysis for problem-solving
C. Quantification

The focus in this discussion is prevention of MSDs. However, nearly the same steps can be taken if the focus is on eliminating inefficiencies and improving production and thru-put.

A. Where to start — Background surveys

The objective here is to analyze your whole facility to get some sense of where the problems are. The problems can be either the worst ones or, as appropriate, the easiest ones to fix, so that you can get experience and build a track record. Obviously, you cannot do everything all at once, so there is a need to set priorities and plan a strategy.

Injury analysis

Review recent injury cases. One of the easiest things to do is to review recent injuries, identify those which are MSDs, and investigate those to determine if improvements can be made. Simply developing a list of these cases is a very good start.

Conduct a study of injury records to find out how many and what kinds of disorders have occurred on which jobs (or in which departments). In the U.S., the primary sources of this information are the OSHA 300 logs and workers' compensation data. The workers' compensation data can have the added benefit of providing *costs* of disorders.

Be aware that in smaller organizations, the overall numbers in these records may be too low from a statistical point of view to make meaningful comparisons between areas. Nonetheless, you need to know what has been reported and where.

Employee surveys

Talk with people about where they have experienced problems and what their needs are. This is the simplest and probably the most effective type of employee surveys. You can talk with people either informally at their own workstations or in group meetings. Or you can attend meetings of supervisors and managers and ask for their input. Usually, you can get a very good sense of issues and opportunities by having these types of discussions.

Solicit complaints and suggestions. By simply talking with people you should get a list of concerns and suggestions, but you can formalize this process a bit by providing a way for people to contact you. Whether they are supervisors or employees, you can let people know that you (or TeamErgo) can come to their work area to do an evaluation. This can be done either in writing (like a suggestion box) or by contacting you directly to the discuss issues.

Use discomfort questionnaires to survey employees. These are usually simple one- or two-page questionnaires that ask people if they experience any physical discomfort from their jobs, usually by referring to a part of the body and an index of severity. The questionnaires are usually anonymous. Most of the time, they also have the benefit of permitting results to be graphed or statistically analyzed. This allows comparisons between areas and "before-and-after" studies.

Often the results of these types of proactive surveys are more accurate and sensitive than an injury/illness analysis, for which the records depend on employee willingness to come forward and report problems.

Conduct satisfaction surveys regarding experience with common pieces of equipment such as chairs or tools and equipment specific to your operations. You can also survey people regarding their experience and perception of the ergo program and solicit input at the same time.

Walkthru survey

Make systematic walkthru evaluations to identify risk factors for MSDs or other ergonomic issues. Often, recognition of some issues emerges only by seeing the task. There may be problems that no one has ever reported.

Look for red flags such as awkward postures, employees wincing when they do certain actions, or makeshift changes made by using items such as masking tape or cardboard.

Identify potential problem areas to follow up with a more complete evaluation. This is the prime purpose for a walkthru evaluation.

You can use a checklist if you wish or just keep notes, even mental notes.

Personnel data

Find out which jobs:
- have high turnover
- are universally disliked
- are lowest on bid lists
- are entry jobs due to undesirability

These undesirable jobs may be prime targets for ergonomic improvement. Sometimes they do not show up on an injury records because people do not stay on these jobs long enough for an MSD to be correctly attributed to them.

Once again, informal discussions with employees and supervisors can be as valuable as a formal study of data.

Medical feedback

Solicit input from MDs, nurses, and physical therapists who have treated employees. These professionals often have unique insights into problem areas based on their own perspectives and conversations they have with supervisors and employees. This input is one of the reasons why it is good to have medical providers on your committee; it closes a loop.

Ideas and solutions

Sometimes you may have a list of problem jobs that no one knows quite how to fix. What triggers action is someone suggesting an idea that might work.

More generally, you should constantly search for problems that you know how to fix. It isn't always best to go after the worst problems, since sometimes the worst problems are very hard to correct.

The point is that in many situations, your "survey" is for suggestions and ideas. Always be on the lookout for good ideas.

Production problems

Look for hectic areas or bottlenecks to see if ergo issues are a problem. These areas may or may not be a source of injuries, but if issues like long reaches and high exertion are causing production problems, you may want to tackle these areas, too.

Renovations and targeted areas

Find out what else is happening at your workplace. Maybe a department is slated for renovation, or a new piece of equipment is being purchased and installed. Or perhaps a Process Improvement Team is targeting a certain area. This may be an opportune time for you and TeamErgo to be involved.

Analysis — To divide a whole into subparts

It is helpful to recognize that the term *analysis* means to take a whole entity and separate it into its constituent parts. (It is the opposite of *synthesis*, which means to combine subparts into a whole.)

This is important because there is considerable confusion on what constitutes *task analysis* — many people think it means *quantification,* which it does not.

To analyze a facility is to divide it into departments and jobs where the worst problems are. To analyze a task is to divide it into its component steps and the principles of ergonomics that come into play at each step. Neither necessarily involves quantification.

B. Task Analysis for Problem-Solving

Once you have focused in on a particular job or work area, you then start concentrating more on the specific steps of a job and the principles of ergonomics that are affected at each step. And most importantly, at every step, think about possible improvements.

Incidentally, don't be misled by this section. If you have a good idea and there's no reason not to go ahead and implement it, go ahead and do so. The following materials are for when you don't know what the problems or solutions are.

Checklists and worksheets

Well-designed worksheets can serve as simple, yet effective, tools. TeamErgo members can readily be trained to identify issues using a good checklist with virtually the same accuracy as a professional ergonomist. Several versions of checklists and checklists are included in this kit.

Be systematic in your evaluation. Focus on each part of the body one at a time (hands, then arms, then back, etc.). A good checklist can serve as a "mind-jogger" of risk factors and can help the evaluator be systematic and break down the task into specific elements.

Talk to employees while you are evaluating the job. Often they will have insights into both problems and possible ideas.

Videotape people at work

One of the best ways to analyze jobs is to videotape employees at work and then review these tapes with a group of people in a conference room. Watching a videotape helps you focus, so that often you will notice things on a video that you do not see right in the work area. You can run the video in slow motion and see movements that you miss in real time. Finally, you can get multiple people involved in a discussion that would be difficult to obtain either observing the job directly or just talking about the job in a conference room without the visual aid. You can even bring in the employees from the area and get their comments and input as they watch the tape.

For these reasons, the video camera is one of the prime tools of TeamErgo as well as the professional ergonomist. In some successful programs, routinely watching videos of jobs is the core of their process. It is effective . . . and inexpensive.

Brainstorm

Get in the habit of brainstorming. You can do this by yourself when observing the job directly, but brainstorming is especially effective when you do it as a group while watching the video. While watching the actions of the person on the videotape, you list all the conceivable options for improvement that you can. Even if you notice little things, if they are easy to implement, you can do so.

It is exciting to see that many problems can be fixed in this way. Sometimes some of the toughest problems have been solved by creative and inexpensive solutions when using this simple approach. Make sure to assign responsibility and action items to keep things moving.

Use teamwork

Teams are good for this type of activity. Different people bring different perspectives, setting the stage for creativity and innovation. Involve some people who are familiar enough with the job so that they can explain what's going on, and involve others who are *not* familiar with the job so that they bring a fresh set of eyes. You can involve engineers and managers who might be familiar with completely different ways of doing the task. You can involve the employees themselves, who may know things that no one else does.

Evaluate results

Make sure there is follow up after any initial changes are made. Often further modifications are needed, either to correct an unanticipated problem or make yet further improvements once you gain experience with the changes. It is all part of the same task analysis process.

Focused survey

Sometimes it is useful to make a special walkthru survey just looking for one item. For example:
- Where are anti-fatigue mats needed?
- What carts need better wheels?
- Where could platforms be used to raise people?

These focused walkthroughs can keep you from being distracted by other issues. Thus, more results can often be gained. This approach is also helpful if you find a solution that works for one of your work areas. If you have similar work areas, you should survey them just for that issue.

Total survey

A good goal is to ultimately evaluate each and every work area and task in a facility. In large site, this effort can take time and resources (years in fact), but it can be worth it. Experience shows that many people who are having difficulties may not report their problems or request help for a variety of reasons.

Moreover, especially in older facilities, people might be performing outdated and inefficient tasks that have all but been forgotten, but yet are still being done. It is not unusual to uncover these types of "hidden" tasks in a total survey.

This is one of the great values of ergonomics. You put on your "ergonomics glasses" and you can *see* activities that you normally walk by without paying much attention.

Increased sophistication

Part of the goal is to increase the analytic capabilities of personnel in the plant. In other words, instead of people merely stating "I *feel* that change '*x*' would be good," we want to help them be able to say, "the *results of our evaluation show* that change '*x*' would be appropriate."

C. Quantification

Sometimes it is helpful (and even necessary) to measure the ergonomics issues that you observe. Examples include: How many arm motions per day are there? How much force is involved? How much time is lost because of bending and reaching?

Be aware that when this style of task analysis is needed, often the goal is *not* necessarily to find ways to improve jobs. Rather, the goal is to describe only what currently exists.

Examples of when quantification can be helpful are:

Measuring physical demands

Conducting "before-and-after" studies — One of the greatest values from measuring physical demands is to document the effects of job changes. Numbers have power and it can be very helpful to show improvements. Some of the success stories in Part I of this book provide good examples.

Matching jobs to employee restrictions — When an employee is under medical restrictions it may be necessary to insure that the physical demands of a particular job do not exceed those restrictions. For example, if a restriction states "no lifting, pushing, or pulling above 20 lbs.," then there is an obvious need to measure every potential action that might occur.

Evaluating appropriateness for job rotation — If job rotation is being considered as an approach to preventing MSDs, it may be necessary to evaluate whether sufficiently different muscle-tendon groups are used. Note that sometimes the differences between tasks are fairly obvious and non-quantitative methods can suffice. But sometimes more rigorous methods of measurement are needed.

Proving that ergonomic issues are present or not — Sometimes it is necessary to prove that a problem exists or that a numerical guideline is being violated. Again, numbers have power and quantitative methods may be needed, especially in reference to regulatory or legal matters. Or, if a solution is costly, there may be a need to document that the problems are excessive and the costs for the improvements are justified.

Conducting research — One of the big challenges in the field of workplace ergonomics is developing numeric guidelines for task design. (How much arm exertion is excessive?) In order to develop these numbers, studies are needed. Obviously, studies of this sort require very rigorous methods to measure the various physical demands.

Documenting background information — Finally, sometimes it is helpful to keep a record of background data on each job.

Advanced problem-solving

Also, there are times when measurements are needed for problem-solving. Examples are:

- When common sense and the simple evaluations described

previously are not sufficient.

- When alternative choices for potential improvements must be evaluated.
- When designing products and precision is needed.

Measurement techniques

Some ergonomic issues are easy to measure, while others are difficult and expensive to do. The companion book to this one (*The Rules of Work*) provides information on these quantitative methods, including:

- Anthropometry (measurement of humans)
- Measuring or estimating force and exertion
- Characterizing working posture
- Counting motions
- Numeric guidelines such as the *NIOSH Lifting Guide* and the *Push/Pull/Carry Tables*

Scoring systems

There are quite a number of task scoring systems that are available, some proprietary and others in the public domain. These scoring systems can be helpful in that they typically enable obtaining a score for each of the MSD risk factors involved for each body part, then combining these results for a total score. In theory, these total scores then enable comparisons between different jobs in order to compare risks.

There are pros and cons for taking this route and the choice depends on your specific objectives. The culture of your organization can also come into play. Some companies encourage numbers for everything; other companies discourage the practice.

Warnings

The greatest sources of confusion in the workplace ergonomics process have been related to task analysis systems.

Confusing your objectives

The approach to conducting a task analysis can differ greatly depending upon your purpose. Failure to make this distinction has been the source of much misdirection and wasted effort.

The most common error is confusing the goal of *solving* a problem vs. that of *documenting* the problem. Documenting problems involves precision in identifying and measuring risk factors. In contrast, problem-solving usually does not.

If your goal is to improve jobs, then quantifying risk factors may, in fact, be detrimental to your efforts. One of the most common mistakes that employers and government agencies make is spending too much time and effort on documenting problems, rather than fixing them.

To improve jobs, it is often sufficient to make qualitative judgments of the issues, and then begin focusing on finding options for improvements. Your time is usually better spent in brainstorming improvements rather than rigorously documenting the problem.

Downsides to scoring

Developing scores for jobs can provide value, if done right and in the context of a good process (and viewed with a grain of salt). The process of developing a score can help you understand the issues involved. However, there are many downsides and potential problems:

- It is not necessary to score a job to fix it.
- Time and energy go into scoring and not thinking about improvements.[*]
- In general, the scoring systems are not yet scientifically valid; there are too many unknowns in how factors interact to be able to come up with a total score.
- Scoring systems set up an all-or-nothing mentality, that you only fix items that exceed a threshold.
- It is very difficult to develop accurate scoring systems, which can result in your whole effort being discredited. The most common situation is where Job A gets a score worse than Job B, but everyone knows that Job B is more difficult.
- When you start developing numerical criteria for tasks, it becomes all too easy to be blinded by the details and stop using common sense and good judgment.

Cautions on checklists

A good checklist can help evaluators to be thorough and provide a framework for discussing issues. But, there can be problems if you don't use them correctly.

It is usually not very helpful to just check a box that indicates there is a problem. You should indicate some type of improvement, even if you aren't exactly sure what it should be. For example, if you check something like "excessive bending," then it is important is to start listing alternatives and ideas for improvement to eliminate the bending.

Consequently, the "blank portion" of the checklist is the most important section – the area where you write down your ideas for making improvements.

Furthermore, it is useful to consider the checklist as the *start* of the problem-solving process, not the *end.* There have been fruitless situations where the TeamErgo members fill out a checklist, pat themselves on the back, and hand off the list to someone else. Usually that is not helpful. Committee members need to be involved in the follow-thru as well. Or conversely, the people who are to be involved in the implementation of the solution need to be involved in the analysis, or kept in the loop somehow.

---

[*] I have seen some workplaces where everyone has become so wrapped up in trying to quantify the risk factors that no one has any energy left over for developing creative ways to make improvements.

Overly complex   There are some worksheets and scoring systems that are so complicated that they are all but unusable. (It's not good when the analysis system is more burdensome than the jobs themselves.)

Long reports   Professional ergonomists and academics have a reputation (deserved) for writing long reports that prove there is a problem, yet are light on ideas and solutions. You don't need to follow that example.*

Finding improvements vs. documenting problems

The most common mistake in worksite task analysis is confusing the goal of finding improvements with that of documenting problems. The latter can require time-consuming measurements that can be detrimental to your efforts if your goal is to solve problems. Finding improvements is usually best done using low-tech methods that do not involve numbers.

Simpler is Often Better

In the practical world of workplace ergonomics, the simplest approaches are often the best:

- Put on your "ergonomics glasses," walk slowly thru work areas, discuss problems with employees and supervisors, and solicit ideas for improvement.
- Take videos of tasks, and then convene a group of people, including the employees themselves, to watch the tapes and discuss issues and ideas.
- Keep the paperwork to a minimum:
  - action items after a meeting: who is responsible for doing what
  - work orders
  - list of improvements made

A good task analysis can be quite simple and yet provide good insight into issues. Don't be overwhelmed by a term like "ergonomic task analysis," which can sound more forbidding than it actually is. The point is to think systematically about each workstation and how each of the principles of ergonomics applies to each part of the body. The analysis is to help you think, and is not the end in itself.

---

* I was fortunate that early in my career I got some valuable advice. From a plant manager, "I don't want a long report. I want a capital acquisition request that I can sign." And from a union leader, "I don't want to hear about all the ways that workers can be hurt. I want a list of things that we can fix."

Making Job Improvements

Finding feasible ways to improve tools and tasks is the *key part* and ultimate goal of the process. Generally, there are three categories of improvements:

    A. Engineering improvements
    B. Work methods
    C. Administrative changes

A. Engineering Improvements

Engineering improvements include things such as buying new tools and machinery as well as modification of existing equipment. Typically, engineering changes are preferred because they tend to get at the root cause of problems, plus they reduce the need to rely on the inconsistency of employee training and habits. Also, when you focus on engineering issues, you increase the likelihood of finding a way of doing the job that that is plainly better from all points of view.

The nature of your challenge can depend considerably on your industry. In some industries, like light manufacturing, every job is different and the challenge is coming up with ideas, brainstorming each workstation. In other industries, the solutions are generally known, and the challenge is to determine what should be used where, often with budget being the primary roadblock. Or there are industries where there are only a handful of problems, but hundreds or thousands of jobs with exactly those same problems. And finally, in some situations, there may be no good engineering solutions at all.

The common thread in many of these situations is the search for ideas for improvement. Usually, this is the biggest question that faces TeamErgo: How do we find feasible ideas that will make a meaningful difference in our jobs?

Sources of ideas

Fortunately, there are many avenues for finding ideas. Some sources specifically target the question of ergonomics. More commonly are good solutions that may not necessarily use the word "ergonomics" at all.

Internet — The best source for information, of course, is the internet. Good ways to search are for:

- "Ergonomics solutions" or "ergonomic products" to find some of the more typical solutions like good chairs, pallet lifts, or adjustable height work tables.
- "Industrial supplies" and "suppliers" to find companies that specialize in all kinds of workplace equipment.
- "Material handling equipment" will narrow your search to companies that specialize in these common types of supplies.

Sources of ideas (continued)

With a little experience doing this sort of thing, you can search with key phrases for almost anything. It is well worth your time (or the time of some other member of TeamErgo) to surf the net in this way.

In-house capabilities — Once the basic concepts of ergonomics are understood and problems identified, almost everyone can have an idea for improvement. There is nothing magic about ergonomics; anything that solves your problem is "ergonomic." Your in-house engineers, managers, and production employees can be the source of many improvements once they are pointed in the right direction.

The many problem-solving tools and analytic techniques that are a part of normal engineering practice are all applicable for ergonomics issues:

- root cause analysis
- cause-and-effect diagrams
- flow charts
- process control charts
- Pareto analysis

Catalogs and trade magazines — There are countless articles, advertisements, and lists of equipment in various publications, most of which are free to you and may already be available on site in your purchasing department. Read:

- vendor catalogs, which contain a wealth of ideas and solutions, especially if you are creative and can think of unconventional uses of standard equipment
- industry trade magazines, which have many articles and advertisements on ergonomics that can be helpful
- trade magazines from *other* industries — material handling, industrial engineering, etc.
- publications on ergonomics, which have more general background ideas for improvement

Remember that a possible solution does not need to be something already labeled "ergonomics." Keep in mind the success story in Part I of this book where we used a potter's wheel to solve a problem in a machine shop.

Field trips — It can be helpful to go elsewhere to get fresh ideas:

- Visit other plants, both in your industry and other industries with good ergonomics programs.
- Attend conferences and trade shows. Many have vendor exhibitions that show the latest equipment, which often means improved ergonomics.

Vendors

If you don't want to search, you can sometimes arrange for the equipment vendors to come to you. There are many knowledgeable sales representatives who are willing to come on site, participate in your meetings on occasion, and support your activities as the circumstances demand.

Build relationships — Most employers have established relationships with reputable vendors, and if yours hasn't, this is a good thing to do. Although their goal is to sell *their* products, many will help you find other items you need as part of an ongoing relationship. They can be a good source of information for all sorts of things.

Explain your needs — A good vendor is always searching for what the market is seeking. A good vendor should want to know about your problems so that they can identify opportunities to provide new products.

Participation — Invite your contractors and equipment suppliers into your meetings and training sessions. They may well be interested in participating in brainstorming sessions, if there is a good possibility of getting the contract for the work that comes out of it.

Specifications — If you use contractors to build and install equipment, make sure that they know of your efforts and that you are expecting equipment that has undergone ergonomic review.

Ergonomics summits

Some companies have taken to hosting "ergonomics summits" where they invite vendors in for a day. The idea is to have all your purchasers, managers, and engineers on hand for one day to be available for consultations with the vendors. You can arrange for members of TeamErgo to serve as hosts to bring vendors on tours thru specific areas of your site and to answer questions. This presents a streamlined way to get bids and proposals from contractors and vendors. It can save a lot of time, both for you and the vendors themselves.

Brainstorming sessions

The importance of holding brainstorming sessions cannot be overemphasized. In companies with successful programs, brainstorming has become a part of day-to-day activities.

Videotape the job in question, then hold a small group meeting in a conference room to review the tape and brainstorm ideas on a flip chart. Be sure to encourage "harebrained" ideas, since they often lead to feasible improvements.

Equipment — If tools are involved, you can brainstorm all the types of tools that are possible, including ones that haven't been invented. You can do the same for every other piece of equipment that is involved, especially mechanical assists of different types

Layout — A common item to brainstorm is the layout of the work area (or desk, or whatever). Do a "what if" exercise: What if Item A were horizontal, not vertical? What if it were placed in a fore-and-aft orientation rather than sideways? Could all similar items be placed together, rather than spaced out? Be systematic and think thru all the different orientations that items can have.

Systems — Take a step back and think about larger issues: Would improvements in the overall material handling system help? Would changes in the overall work process help? And most importantly: Is there a completely different way of doing the job?

Mindset questions — There are a number of questions that you can ask yourself that can help you get in a brainstorming mindset:

- If you were a Yankee inventor living in 1820 and had no electricity or power, how would you do this job?
- If you had unlimited resources, what would you do?
- If you had *no* resources, what would you do?
- Have you ever seen a different way of doing this task? What implications does that have in this case?
- Does a similar task exist in another industry? How do they do it there? What are the implications for you?

Adapt from automation — A valuable thought process is to review any automation that exists for the task you are trying to fix. Perhaps the automation is too expensive or not feasible in this case, but there might be some aspect of that automation that you could use. Remember the example from Part I where a flipping device was adapted from an automatic printer to a manual one that saved both time and motions.

Implementation

Once you have some good ideas, implementing them should be straight forward, assuming you have the budget and the necessary approvals. The following are some suggestions that can help you in this process.

Trial and error — Ergonomics improvements are not always straightforward and a period of experimentation is often necessary to find a good improvement. A good example is work height, where the proper work height is determined by a combination of posture, strength, and visual needs, all of which are very task specific. You can't just look in a book to find out the proper height.

The concept of continuous improvement is important here — a job improvement is planned, then implemented, then evaluated, and then refined in an ongoing process.

Maintenance staff — A stumbling block in many ergonomics programs is a shortage of maintenance staff to implement the improvements that have been identified. Consequently, in some successful programs, the company has added maintenance personnel and/or assigned maintenance staff to the ergonomics project. In some cases, companies have increased the use of outside contractors, since it provides more temporary flexibility.

Ergo work orders — Some larger companies with significant MSD problems have instituted special work orders that give ergonomics projects priorities. Using a "safety" designation for these ergo projects would also serve the same purpose in many companies.

Long-range planning — Some projects can only be achieved in the long term, but you should do your planning in the short term. Don't wait until construction begins or the purchase order is sent before you think about what you actually need. Start making your lists now.

Time frame

Not all changes can be implemented at once (or in some cases *should* be). Thus, you should think about improvements in three general categories: quick fixes, simple solutions, and long-term renovation. It can be especially important to provide feedback to employees if the changes you intend are going to take some time. Most of the time, people would rather hear that it will take some time than not hear anything at all. (And most people would rather hear "no" than hear nothing at all.)

Quick fixes you can do today or tomorrow. Examples are (1) layout changes to improve heights and reaches and (2) improvements in employee work methods.

Simple solutions are more permanent and can be done in a month to several weeks. Typically this category includes equipment that takes budgets and special approvals, or involves time to process an order and ship.

Long-term renovation is generally more costly and requires additional time. Sometimes you need to wait until a major renovation is scheduled before some improvements are feasible. What is important at present, however, is to keep good notes of what your needs are. As mentioned previously, don't wait until the construction crews arrive before you start trying to list all the changes you've wanted to make.

Ergo log

Make sure to keep a record of the projects you are working on and the improvements you have made. This does not need to be complex or difficult, and it certainly helps you to keep track of the status of everything. The system can range from a simple notebook to a computer spreadsheet or data base system. (See the materials on Ergo logs in later sections of this book.)

Follow-up

Make sure to check up periodically on the changes you have made to see if they are working the way you have intended. You may need to do some fine tuning or even go to "Plan B" if the change is not as effective as you expected.

In this respect, ergonomics and reducing MSD risk factors is no different than other aspects of workplace design — there may not be any one permanent solution, rather there will be opportunity for ongoing and continuous improvement. In this context it is quite appropriate to make small changes now, with the expectation that larger and better changes will be made later.

Communicate

Be sure to keep affected people in the loop. Especially for the employees whose work areas are being changed, it is simply polite to let them know. Furthermore, the more you involve people, the more they are likely to accept the changes and cooperate with you in implementation and refinement of the ideas.

Keep in mind that it may be helpful to communicate the results of a change in one area of your facility to employees elsewhere in the facility. This helps keep a sense of momentum going thruout the entire site, even if you are only focusing on one or two areas. Also, before-and-after information of this type helps to educate and inspire people.

Finally, remember to give credit to originators of ideas and teams that have worked on projects.

Hire experts in ergonomics

Professional ergonomists can be a useful part of this process. The ergonomist can provide a new set of eyes to see issues in the workplace that you may have become accustomed to, or to double-check that you are on the right track. Finally, the ergonomist may have experience with improvements in other industries and workplaces that may be of value to you.

B. Work Methods

The best improvements are usually engineering controls, as described above. But sometimes mechanical fixes are not possible and the focus lies with improving work methods. This can be a difficult challenge since it involves training every employee and perhaps trying to change ingrained habits.

As stated previously, it is not uncommon to see two employees working side by side, doing the same job, one person smoothly and calmly while the other person appears hectic and awkward. Typically, the person working smoothly is more productive and at less risk for MSDs. The issue with work methods is how to train that hectic second person in the smooth methods of the first one.

Work methods can be divided into two categories, task-specific methods and general rules that can apply anywhere.

Task-specific

We are all familiar with tasks that require methods that are very specific to that task. In fact, it is difficult to think of any type of physical labor that doesn't have this component.

A good example is moving furniture. First of all, finding

engineering controls is all but impossible, at least beyond the traditional use of dollies and hand trucks. (Whoever invents a feasible device to move a sofa bed around a corner on a stairs should become rich.)

An experienced mover is a master at techniques and little tricks to make the moving easier. These skills make the job easier, with less risk of musculoskeletal injury, but it may be difficult to formalize these skills into training lessons.

From a training perspective, we can divide this type of problem into two categories: knowledge and enforcement.

Knowledge issues — Sometimes the question is how to capture and describe the skills and knowledge that an experienced employee has gained over a long period of time. Furthermore, we ask ourselves if it is even possible to "train" people in these subtle methods, other than letting them slowly learn by themselves, or perhaps by working together with the experienced employee.

Enforcement issues — The other situation involving task-specific work methods is when we know the best method and it is easy to train people in that method. However, we need to *enforce* the use of that method.

The job of picking orders in a distribution center provides a good example. Again, finding mechanical assists for this task has proven difficult — at some point the employee has to pick up a box and put it on a cart or pallet jack.

There are many methods that affect this task, but probably the best instance of a task-specific rule here is "picking in layers." The challenge we face is that everyone must do it. The temptation is pick the closest box to you, whether on the right layer or not. That makes it easier for you at this point, but more difficult overall.

In short, we know the rule and it is easy to teach, but harder to ingrain it into everyday work habits.

General rules

There have been quite a number of general rules that have developed thru the years.

Lifting technique — The classic example is good lifting technique: keep the load close; maintain the arch in the lower back; move smoothly; check the weight before you lift; etc. These methods can be applied in a multitude of situations.

Examples of other rules — There are any number of rules of thumb along these lines. Examples include the following:

- Take a step rather than reaching or twisting.
- Let the tool to the work.
- Push rather than pull.
- Adjust equipment or workstation prior to use.
- Use the non-dominant hand instead of reaching across the body.

- Don't stack items too high.
- Avoid unnecessary exertion (for example, avoid gripping hand tools too tightly, or pounding on a computer keyboard instead of typing lightly).

Approaches to training

Identify the best method — In those cases where knowledge of the best method is the roadblock, it can take you a fair amount of effort to discover the best technique, if any.

The simplest approach is to videotape several employees, then hold an employee meeting to watch the tape and discuss the issues involved. (Note that this situation must be handled carefully so as not to inadvertently hurt anyone's feelings. Watching yourself on video is usually humiliating enough, let alone being critiqued at the same time. Sincerity and sensitivity on the part of the meeting facilitator can go a long way here.)

Another, more scientific approach is to study the same videos of employees working, using slow motion and pausing the action to analyze specific steps. In some tasks, both good and bad methods can be immediately observable. In other tasks, quantitative job analysis may be needed to differentiate between different work methods.

Transferring Information — Once you know the best method, you need to tell people about it. Sometimes this can be done simply, such as in a booklet or new-hire orientation sessions. At other times, it might involve meetings like those described above where employees watch videotapes of good and bad practices. Or you may find it useful to video your best employee using the best practices.

Note that watching videotapes to learn subtle methods often *cannot* be done in a new-hire orientation session. The employee is too new and still unfamiliar with the job to grasp the subtlety. On the other hand, you don't want to wait too long either, for fear that bad habits will develop.

Another comment is that this approach of watching videotapes of employees doing the work has not been that common. It is a new concept related to the recent availability of inexpensive, high-quality video cameras that opens up the opportunity.

Practice — It is not enough to just be told *how* to do the task. You must *practice* for a period of time to learn. Thus, to train people in good work methods, you must build in practice time.

Follow-up — The trainer needs to be involved thruout the practice period to monitor the learner's technique, demonstrate the better method as often as necessary, and stay involved until the techniques become an automatic habit.

Changing behaviors

Everyone on TeamErgo should be aware that usually it takes a while to change peoples' habits. Some individuals will adopt everything you say and are appreciative, but others will take time, and may never do what you think is best.

Also, be aware that it is common for people to become so used to working in a certain way that *any* change, even if clearly beneficial, can seem awkward and uncomfortable. Sometimes you need to cajole employees into trying a new tool or technique for a week or so before they become used to the change.

C. Administrative Changes

The third category of job improvements is changing various types of administrative practices and management systems. This category is also sometimes called *work organization* and can involve a very broad range of activities, ranging from minute issues like task allocation (which person does what) to almost philosophical issues like using self-managed work groups instead of traditional hierarchical line management.

Administrative changes can be implemented for a wide variety of reasons, including increased efficiency, increased employee satisfactory, or reduced risk of MSDs. Some of the more common practices related to MSDs are outlined below.

Job rotation

Perhaps the most common administrative change sought for reducing of MSDs is job rotation. There are many reasons for implementing a job rotation system. Employees can become more knowledgeable about different jobs and thus be in a position to offer better input about process improvements. Production can become more flexible because of the cross training. Employee satisfaction can also increase. And, under the right circumstances, rotating jobs can help prevent MSDs.

Different muscle-tendon groups — It is crucial to recognize that job rotation done for the purposes of preventing MSDs must be done in a way that alternates use of different muscle-tendon groups. It does no good (from an MSD perspective) to switch people among tasks that are substantially the same.

In some facilities, finding suitable jobs is easy and there may be a wrist-intensive job immediately adjacent to a back-intensive one. Indeed, sometimes employees spontaneously switch tasks to divide up the type of work they are doing.

Job rotation plans can be very simple, such as alternating people at a single task that is particularly burdensome. Or they can be complex, involving everyone in the facility.

Caveats — The main problem with job rotation (from an MSD perspective) is that it doesn't do anything about the risk factors present in a facility. It only distributes the risk factors more evenly across a larger group of people. Thus, the risk for some individuals will be reduced, while the risk for others will be increased.

Job rotation can be seductive, in that it seems like an easy solution: "We can just rotate people around and then not have to worry about task analysis and having to think about improving our jobs." Consequently, many job rotation schemes are instituted without much thought and don't have much impact, or may even make things worse.

In fact, setting up a facility-wide system of rotation can be exceedingly difficult. There are many issues to consider and a multi-staged implementation process is often necessary. (More background information and a set of guidelines for establishing a job rotation system are found in a later chapter.)

Job enlargement

The concept of job enlargement is to increase the scope and duties in jobs and thereby broaden the division of labor. The classical examples come from the auto industry, where an assembler might install multiple components rather than a single one, plus be involved in decision-making for the operation.

The primary reason for job enlargement is to increase job satisfaction and product quality. Employees tend to become more familiar with different aspects of the work and thus more able to spot inefficiencies and inferior procedures. Additionally, in many cases, there can be sufficient variety in activities so that no one part of the body becomes overloaded.

Cycle time is inherently increased with enlargement. Instead of having only 12 seconds to install one component, the employee might have 80 seconds to install a sequence of parts. Longer cycle times tend to give employees a bit more flexibility and thus remove some of the stress of the jobs. Furthermore, the longer cycle sometimes enables use of tools and equipment that are not feasible with short cycles and that can have ergonomics benefits.

Perhaps the most important benefits are the philosophical and psychological aspects of changing the perception of one's job. Instead of being an assembler of component "3212y," now the employee can become a "co-manager" of the work area.

The system (if designed this way) thus helps to tap into everyone's capabilities, engaging the whole person, *not* just the hands or the back. Job enlargement helps to get rid of the dead-end nature of the extremely narrow division of labor.

Pay systems

Reward systems can also be part of administrative changes to reduce risk for MSDs. For example, old-fashioned piece rate systems tend to encourage people to produce as fast as possible to the detriment of product quality, their own well-being, and effectiveness of the whole work process. Better systems include rewards for improvement of the whole work process or pay for knowledge and skills.

This topic is too complex to be discussed here. The point is to remember to think about reward systems.

Overtime

Excessive overtime increases the risk for MSDs, since the exposure to MSD risk factors like force and motions are increased, while simultaneously, rest and recovery time is reduced. Overtime can also negatively affect productivity, quality, morale, and accidents. Furthermore, excessive overtime probably contributes to sleep deprivation, which itself can lead to serious injuries, such as in vehicle accidents.

As appropriate, you may need to curtail excessive hours to reduce exposure to MSD risk factors and increase recovery time.

Stretching and exercise

Various approaches to stretching and exercise also fit into the picture. There are three general categories:

Energy breaks are designed primarily for highly sedentary jobs where you stop work from time to time to stretch and get your batteries recharged. Often employees can do this sort of thing individually and informally, such as by stretching back in a chair or squatting for a few seconds in a standing job.

Energy breaks can also be formal and led either by a trained fitness instructor or a volunteer. Regularly scheduled breaks like this are not uncommon now in some companies and, in some places, exercise breaks are even mandatory.

Warm-ups are targeted for strenuous work, where employees warm up a bit prior to beginning work. A good example is a maintenance worker in a steel mill who may spend time in an air conditioned room, then be required to perform heavy exertion. A bit of a warm-up can be very helpful in situations like this.

Industrial athlete programs are aimed at improving the physical fitness of employees, not just stretching or warming up. Implementing a system of this type is an ambitious project and is often easier to achieve if there is an active wellness program.

Medical Management

Good medical systems are an essential part of the MSD reduction process. Altho this is a relatively new area of medicine, it is evolving rapidly. There are still topics where medical opinions vary, but there is a growing consensus on the broad categories of activities that should be addressed.

The following section is written from a layperson's perspective, that is, there is nothing here on professional-level diagnosis or treatment. Rather, the focus is on the types of things that a member of TeamErgo should know.

Early recognition

The most important thing for you to know (as mentioned in Chapter 3 *Understanding MSDs*), the sooner that symptoms are recognized and reported, the better the chances are for effective and inexpensive treatment. Consequently, it is important to set up procedures to find employees who are experiencing problems as soon as possible, before lost days and surgeries occur.

The primary methods for accomplishing this are training sessions and/or employee handbooks in which everyone is told to report problems as soon as they occur. Other mechanisms such as workplace newsletters and bulletins can also be used for this purpose.

Why MSDs are different

Most guidelines or standards for occupational disorders contain sections on medical issues, such as the requirement for hearing tests when employees are exposed to noise. The guidelines for the prevention of MSDs are basically no different.

There is one important aspect of MSDs that makes these disorders fundamentally different from many other occupational health hazards. This difference can perhaps be best described by contrasting musculoskeletal disorders with hearing loss.

The audiometric exam for people exposed to noise is a common screening test. Unfortunately, once you are diagnosed with hearing loss, there is not much that you can do about it. There is no surgery or drugs that you can take to bring your hearing back.

Treatable

MSDs, however, are *treatable*. If the MSD can be identified sufficiently early, there is a great deal which can be done for treatment. If detected early, many MSDs can be treated with simple measures such as anti-inflammatory drugs, cold packs, and restricted duty. Even if diagnosis is delayed, there are still methods of treatment, such as surgery.

Consequently, it changes the purpose of getting a medical exam and sets up a strategy that is often not possible with other health hazards. This is where TeamErgo comes in. You need to help people understand that they should report problems rather than trying to tough it out.

Taking this step challenges a lot of traditions. However, it is worth it since you can keep the little things from becoming big things. The first time you learn that someone is having some problems ought not be when they are being scheduled for surgery.

Diagnostic tests

There is another way in that an MSD is different from hearing loss. The diagnostic test for hearing loss — the audiometric exam — is common, well-accepted, and fairly objective. But for MSDs, there is no single diagnostic test. Instead, the methods differ substantially depending upon which part of the anatomy is affected. Furthermore, these types of tests have not been very well known in the medical professions, at least until recently. Finally, and most importantly, the tests themselves are not as objective as diagnostic techniques for many other or disorders. One of the most common symptoms for musculoskeletal disorders is pain, which cannot yet be measured or determined independently of employee statements or complaints.

Implications

Consequently, these factors create opportunities that don't happen for many other occupational hazards. You can develop a strategy for recognizing MSDs early and treating them conservatively, that is, with ice packs, rest, and over-the-counter anti-inflammatory drugs.

An additional factor has to do with the fact that MSDs are very common and affect almost everyone. As long as we do manual activities (whether at work, at home, or in sports or other recreational activities), there is a potential for experiencing MSDs, at least in mild forms. We can't prevent all cases. But we *can* keep the mild cases from becoming major ones. Once again, this speaks to developing a strategy for recognizing problems early and treating them in early stages while it is still possible.

Actions

Review your current medical system to ensure that your medical providers — whether in-house or contracted — have sufficient capabilities and training to provide medical services for MSDs.

Work closely with your medical providers:

- Involve medical providers as members of TeamErgo.
- Invite contracted medical providers to your site to review work areas and learn about your program. Encourage them to keep abreast of MSD issues.
- Involve in-house medical staff in periodic workstation analyses and walkthroughs.
- Solicit input and insights on what medical providers see as problem areas.

Do your best to determine if your medical providers are using standardized diagnostic tests when employees report symptoms and standardized procedures for treatment and referrals.

Develop a referral network of medical specialists for MSDs.

Work with medical providers to ensure that treatment procedures are conservative. The primary goal is to treat any health problem at an early stage with simple treatments (rest, non-prescription anti-inflammatory drugs, etc.) and avoid surgery or other more complex treatments.

Make sure that employees:

- return to work only when deemed ready by the treating physician
- are put on jobs that are compatible with their restrictions
- are evaluated periodically to see that problems are not recurring

Proactive approach

Consider providing physical examinations for employees in high-risk jobs (sometimes this approach is referred to as "active surveillance"). In other words, go out and actively seek employees with problems, rather than wait from them to report. Altho this approach might sound like you are deliberately asking for trouble, it can in the long run save the employer money. You might structure the exams to be more or less like part of a wellness program and address *all* MSDs, whether work-related or not. There are some complexities here, but they should be resolvable.

This step is the equivalent of providing periodic medical exams for other occupational hazards, such as annual audiometric exams for people exposed to noise. Although providing proactive exams for symptoms of MSDs is not a widespread practice, it may be beneficial to initiate such programs, at least on a pilot basis.

MSD case investigation

Investigate MSD cases just as you would a workplace accident. Find out:

- What is the diagnosis? (Some details may be considered private medical information and not available, but basic information is important to have and should be available to everyone on your team.)
- What job factors might be related? (Note: for the purposes of this investigation, you are *not* necessarily trying to make a final determination if the injury is work-related or not.)
  - What possible risk factors do you see?
  - What activities does the employee think might have contributed to the problem?
- What aspects of the medical management system have been applied to this person?
- What have been the outcomes?
  - What changes were made to the task?
  - What happened to the person?

Restricted duty

The system surrounding how an employer finds appropriate work for individuals with restrictions is important, but is a topic that is beyond the scope of this kit. However, a few general points are:

- Current medical practice is to return patients to normal activities as soon as possible. In the work setting, the protocol thus is to return the employee back to work quickly, but not risk re-injury.
- It is important to have good communications with the medical provider to accurately understand what an employee's restrictions are.
- It can take some time and effort to find appropriate work for someone with restrictions (and sometimes it is not possible). In larger organizations, there should be procedures in place for this purpose.

Recordkeeping

It is essential to keep accurate records of employees who report symptoms. The primary form for this purpose in the U.S. is the OSHA 300 Log. Every site should review its procedures for entering information on this log to ensure information is complete and accurate.

Effect on injury rates

As described previously, it is very likely when you institute a good medical management system that your *recordable* injury rate may actually rise, at least for a while. This can be a good thing, since it usually means that you are identifying problems at an earlier stage than they would have been otherwise, which as we have discussed, can be beneficial to both the employer and the employee. Your lost day cases and surgery cases should drop almost immediately, but the less serious recordable cases may go up because of the better reporting system.

You should make sure your senior management group is aware of this likelihood, so that they are not surprised. You may also want to change your bonus and reward systems so that they are not negatively affected by reports of MSD symptoms. MSDs should be treated differently from other types of injuries, for example, those accidents caused by inattention to safety rules.

Another implication is that recordable injury rates cannot be always be used as the sole measure of how safe a particular workplace is. These statistics are almost totally dependent on employee awareness. Consequently, sometimes employers with the best systems and most attention given to safety have high employee awareness, and thus, misleadingly "high" rates of recordable injuries.

Sometimes just hiring an occupational nurse raises the rates. With the nurse available, it is easier to report problems and get treatment, which means the recordable rates go up.

Medical issues

A layperson does not necessarily need to know much about medical evaluation and treatment. However, members of TeamErgo should know that the medical management programs for MSDs can be complex. Special systems must be developed for both the ambiguities and the varieties of problems:

Surveillance — How do we find out who has problems?

Evaluation — By what criteria? And for which disorders: the lower back, shoulder, elbow, wrist?

Referral — When can issues be treated at the plant level? And when should they be referred to specialists? By what criteria?

Treatment — What treatments are effective? Especially, what treatments in early stages? For example, when should cold be applied, and when should heat? When should the patient exercise and when not? And what type of exercise? And for what part of the anatomy?

Return to work — When can the employee return to work? Can the employee return to normal work? Or is restricted duty needed? What kinds of tasks are appropriate for restricted duty? For how long? How do we check employees to make sure they are not re-injured?

Case management vs. medical management

There is mixed terminology in the medical field regarding the difference between case management and medical management. In general, a good way to differentiate the terms is the following:

Medical management is what the health care professionals do to make a person healthy again.

Case management is what non-professionals in the workplace do to follow thru on that person's case.

Ergonomics vs. MSD prevention

One final comment: Technically, medical issues are not part of ergonomics. Common usage in the U.S. is sloppy on this point. Because the field of ergonomics has proven to be effective in reducing MSDs, there has been a tendency to use the term "ergonomics program" as a synonym for "MSD prevention program," which is technically not the case.

There are many aspects of the field of ergonomics that have nothing to do with MSD prevention, for example, cognitive ergonomics or using the tools of ergonomics to improve efficiency. Simultaneously, there are many aspects of MSD prevention that clearly are not part of ergonomics, such as wellness or treatment programs.

A good medical management system should dramatically reduce lost day MSDs and eliminate surgeries entirely. If employees are undergoing surgery for work-related MSDs, you probably need to improve your system.

Monitoring Progress

The overall program should be evaluated periodically. There are a number of objectives in doing so. In part, you want to take a step back from time to time and see if there is anything that you need to improve. Also, as your process matures, you will want to consider if it is time to change what you are doing (in other words, you would do something different in the early stages of a program than in later stages).

And finally, you would usually want to create a good record of what you have accomplished. Typically, you do more than what you remember, so keeping some account of activity will help keep your momentum going as well as communicate to others what your activities have been.

Ergo log

One of the easiest and most effective things to do is keep a list of all workstation improvements. This log may include items such as department, ergonomic issue involved, a description of the improvement, cost, date, etc.

An electronic sample of an Ergo log is part of this kit and is described in more detail in a later chapter. You can use this log as a basis for reports, newsletter articles, presentations, and as a reminder to you that you are indeed achieving progress. And if you are ever visited by a compliance officer, it would be one of the first things to show.

Internal website

You can also maintain the Ergo log on an internal website. There are many advantages of keeping your records this way.

- It provides an easy way for you to update your information in a manner that keeps everyone else current too.
- You can add digital photos, in particular, before-and-after photos of job modifications that you have made.
- You can link each item on your Ergo log to a page that contains more information. Thus, your log can be list of single line items, each linked to a more elaborate description.
- All policies, plans, forms, and updates can be kept in one spot that is easily accessible by anyone with a computer on your network.

Management review

Another simple activity is holding management reviews of your overall activities on a regular basis to ensure that your goals and objectives are met. This kind of activity happens in many organizations automatically. You can provide results in oral and written presentations and discuss both the successes and the problems you are having.

Injury and cost trends

You can track trends in MSDs by graphing your monthly and annual injury cases. Many organizations already generate graphs of overall work-related injuries and illnesses, but the idea here is to break out the MSDs from the overall incidents and track them separately.

Similarly, an important metric is workers' comp costs trends. Tracking the portion of these costs that is related to MSDs can be a lot of work, but it usually worth it.

Before-and-after studies

You may also wish to consider conducting before-and-after studies of job improvements. These evaluations can include items such as:

- Changes in risk factors (force levels, number of motions, posture improvements, etc.)
- Comparisons with numeric guidelines or standards
- Efficiency measures (output, time savings, scrap, etc.)
- Employee satisfaction, either from surveys or individual statements.

A number of examples of these types of studies were provided in the success stories in Chapter 4 *Case Studies*.

Feedback

Employee feedback can be an important measure of success. Surveys of employee satisfaction are valuable, both of small numbers of people affected by a particular change or of the entire workforce. Subjective statements can be important, too. I am familiar with one situation when a single upbeat comment by a hardnosed union representative that "the attitude on the work floor is completely different" changed the view of an equally hardnosed corporate executive from being negative toward the program to being supportive.

Other metrics

All of the above serve as methods to evaluate your project. Additional metrics can include:

- Changes in turnover and absenteeism
- Changes in sickness leaves
- Number of tasks evaluated
- The number of people affected by improvements
- The number of training sessions held
- Time spent by personnel on ergonomics projects
- Compliance with good work methods
- Audit scores
- Safety grievances (or overall number of grievances)

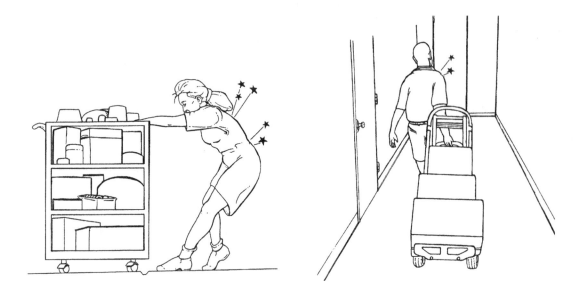

In the past, much of the focus of ergonomics has been in manufacturing and office work. Now the service industry is coming into the prominence for development of ergonomics programs. However, the process for identifying and making improvements remains much the same.

Chapter 6
The Workplace Ergonomics Process

The previous chapter described *program* elements. This chapter focuses on *process*, that is, the stream of actions that bring about a result.* We will first review the process of solving problems, which is the ultimate goal of the ergonomics effort. There are many ways of characterizing this process, several of which are outlined below to help you, depending upon your own style and organizational culture, or simply to present ideas in different ways. Then, the topic changes to the process of introducing ergonomics to your organization and the process of running the program.

Problem-Solving Process

The generic process follows these steps. You can apply these steps as the situation demands to be done simply or more elaborately.

Identify
Evaluate
Improve
Re-evaluate

These steps are no different from normal continuous improvement process of Plan, Do, Check, and Act. The terms and sequence are just modified slightly to fit the ergonomics parameters.

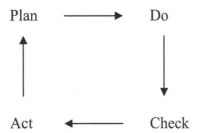

Plan ⟶ Do

Act ⟵ Check

If you like acronyms, you could also use SAFE. This slant is catchy, but de-emphasizes the efficiency aspects of ergonomics and the point of enhancing employee capabilities.

Survey
Analyze
Fix
Evaluate

---

* Often in this book, these two terms have been used interchangeably. Here they are intended to have different meanings.

Simple Approach

Some issues are self-evident and easy to resolve. There is no need for any bureaucratic complexity: no measurements, no formal meetings, no fuss, no muss.

Identify

1. Identify a possible ergonomics issue.

Evaluate

2. Obtain input from affected parties.

3. Photograph the task beforehand.

Improve

4. Arrange to make the needed change.

5. Communicate with affected parties that the change is being made.

Re-evaluate

6. Check later to see if the improvement is performing the way intended or if it needs refinement.

7. Take photographs afterward.

8. Update the item in the Ergo log.

A More Elaborate Approach

At other times, there are a lot of issues and interactions with other processes within the facility, plus there may some investment costs to justify. In this case the process may become more elaborate. One approach is outlined below, which you would undoubtedly need to customize to fit your situation. The bulleted items associated with each step represent alternative ways to accomplish that step.

Identify

1. Identify a possible ergonomics issue.
   - Solicit ideas or reports of problems.
   - Review recent injuries.
   - Conduct an analysis of injury and illness records.
   - Conduct task evaluations, ranging from informal walkthrough surveys to formal studies.
   - Hold meetings with employees and/or managers and supervisors.
   - Solicit input from medical providers who have seen employees.
   - Review "inactive" ideas in the Ergo log.
   - Review trade journals and ergonomics journals.

2. Bring to a TeamErgo meeting (or elsewhere, as appropriate).
   - Determine how to handle this idea (e.g., act on imme-

diately, gather more information, or put on hold).
- As needed, assign a team to follow thru.
- As needed, refer the issue to another group or person.
- Prioritize compared to other issues.
  - Evaluate injury statistics for this work area or task.
  - Estimate rough costs for possible improvements (i.e., is this a $100 project, a $3000 one, or over $10,000? or other similarly broad categories).

3. Enter the issue in the Ergo log.

Evaluate

4. Clearly define the issue and the cause.
- Put in writing what you know.
- Study the job using a worksheet.

5. Obtain employee and supervisor input.
- Hold one-on-one discussions on the plant floor.
- Hold small team meetings to discuss the idea.
- Hold meetings with all the employees.

6. Photograph or video the task.

7. Determine if additional background information is needed or not. If so, obtain measurements:
- Force gauge
- Lifting Guide
- Anthropometry tables
- Scoring system
- Other methods

8. Determine if (a) the solution is obvious or not and (b) if there is consensus that it is a good idea. (These steps could be done by a few people or in a TeamErgo meeting.) If not:
- Hold a brainstorming meeting, including employees and supervisors in the area and members of the Ergo Team.
- Check equipment catalogs and websites.
- Check internet vendors, for example by using search engines and industrial supply companies.
- Use other problem-solving tools, such as fishbone analysis.

9. Get appropriate approvals.
- Handle verbally in a TeamErgo meeting.
- Prepare a background document and presentation to show the results of evaluations, measurements, discussions, and final recommendations.

Improve

10. Go to the appropriate function to make the change.
    - Handle verbally in a TeamErgo meeting.
    - Contact maintenance, operations, engineering, purchasing, etc. as appropriate.

11. Communicate to other employees and managers (as appropriate) that the change is being made.

Re-evaluate

12. Check later to see if the improvement is performing the way intended or if it needs refinement.
    - Employee surveys and discussions
    - Before-and-after measurements

13. Take photographs or videos.

14. Update the item in the Ergo log.

15. Prepare information to be used for updates at various meetings, service talks, bulletin boards, etc.

Comments

These two models should give you a good sense of the ends of the spectrum. You should be able to see how you can set up a process that's right for you.

Note that both of the above processes are described from the perspective of the TeamErgo, but there is nothing to prevent anyone such as a supervisor, engineer, or maintenance staff member from taking initiative in their own areas of responsibility. On the following pages are two different descriptions of the same process written from a more general perspective.

As indicated previously, the reason for providing these various versions of the process is to help you find the one that fits your style and needs the most closely. Also, you may find that the different approaches can be used at different times or settings.

The Problem-Solving Process

1. Identify priority tasks

Where have most sprains and strains type injuries occurred?

What tasks are the most strenuous or frustrating?

What are the tasks that people dislike doing the most?

What tasks occur most frequently?

What are most crucial to accomplish quickly?

What are most crucial to accomplish correctly?

What tasks would be most easy to improve?

2. Fact-finding

Evaluate the task using an ergonomics worksheet.

Videotape the task.

Identify the important ergonomic issues.

3. Brainstorm

While observing the task, think of creative ways to make improvements.

Meet in a conference room, watch the video, and review the worksheet.

Brainstorm possible ideas; harebrained ideas are required.

4. Develop action plans

What ideas are feasible?

Any that can be implemented immediately?

What additional information do you need to gather? Who will do this?

What types of long-term changes are possible?

5. Communicate

Make sure the ergonomics team and other appropriate individuals are updated.

People Watching

Sometimes you look at the tools and equipment to spot problems. But most often you simply watch people.

Step 1— Observe

Observe employees' work methods and behaviors for ergonomic issues.

Postures
- bent or twisted back
- bent neck
- bent wrists
- elbows away from the body
- outstretched arms

High Exertion
- people straining while doing a task
- tense tendons (especially in hand)
- obviously fatigued

Method and Motions
- hectic-looking activity
- jerky motions
- fumbling for materials
- double-handling
- time-consuming manual activities

Static Postures
- standing one foot on top of another
- sitting in unordinary postures
- fidgeting

Discomfort
- wincing — pain or overexertion
- shaking out their hands — trying to alleviate numbness
- rubbing — in particular elbows and wrists
- blisters, calluses, red marks, bruises — contact stress or lack of clearance

Heights and Reaches
- which rules apply? (remember exceptions to the rules)
  - elbow height rule
  - shoulder and knee height rule
  - preference for work surface changes (rather than standing platforms)
- exceed reach envelope?
  - *forearm* reach for *continuously* used items
  - *full arm* length reach for *routinely* used items

Step 2 — Ask Yourself: Why?

If you see these behaviors, now ask yourself why. What's behind the problem? What's the cause of the behavior?

Layout or Equipment Problems?

- Leaning, bending, and twisting to:
  - compensate for poor layout?
  - avoid glare?
  - avoid sharp edges or obstructions?

- Bent wrists because of:
  - inadequate tool design?
  - poor orientation of equipment and/or products?

- High exertion because of:
  - poor leverage on tools and equipment?
  - awkward positioning?
  - small wheels on carts, dollies, etc.?

- Hectic work because of:
  - poor layout?
  - equipment not set up right?

- Static load because of need for:
  - foot rests?
  - sit-stand workstations?
  - adjustable furniture?

Training Problems?
- adjusting chairs and equipment properly?
- proper steps of job?
- best work methods?

Old Habits?
- need feedback?
- need retraining and/or reinforcement?
- need time and encouragement to learn better methods?

Step 3 — Start Problem-Solving

At this point you're doing basic ergonomic task analysis and problem-solving. Start brainstorming improvements

Special applications
The above descriptions should fit most workplace operations, one way or another. The following are some common variations on the general theme.

Light manufacturing — The process may involve a review of each and every workstation with all the tools, equipment, and material handling issues involved at each. This can be very time intensive and may take several years to evaluate every workstation. However, the solutions tend to be less expensive, take less work to justify financially, and can be implemented as you go. The process takes much more involvement of supervisors and employees, since they may be the only ones who know enough specifics about the task to be able to identify solutions that work. Brainstorming is especially important in these situations (and for which I have special appreciation because it is fun and satisfying to come up with very simple improvements that make the job better from all perspectives.)

Process industries — In many of these instances, there are only a few tasks that involve problems, but these problems are repeated everywhere and affect multitudes of people. Thus making improvements may be expensive and require considerable study and cost justification. Fewer people are typically involved and they tend to have higher levels of expertise, such as engineers and architects.

Maintenance — The process of making improvements in maintenance tasks is different primarily because the tasks are often done only sporadically. For example, some maintenance tasks are performed only once every several months (or years), and then done at night or on a weekend. Consequently, it is difficult for TeamErgo or other experts to observe the task and understand what is involved. (This is in marked contrast to many production jobs, which are often performed constantly and available for observation at most any time.)

Thus, with maintenance tasks, you need to plan ahead. You or other members of TeamErgo may need to accompany the maintenance personnel at an odd hour of the night. Or you may need to arrange to have the maintenance personnel themselves videotape the tasks, plus provide them with more training on principles of ergonomics.

On the positive side, the maintenance personnel are often the ones who can make the changes. Furthermore, many issues can be resolved very easily just by getting these personnel to think about what they're doing, rather than take it as a given that the task must be performed in a certain way.

Special applications (continued)

Health care facilities — The ergonomics issues in hospitals, nursing homes, and other health care facilities can be divided into two categories: (1) lifting patients, and (2) everything else.

Lifting patients is generally the most high-risk task in these settings, but a difficult one to resolve. Fortunately, there are many new types of lifting aids that have come on the market to help in these situations, plus hospital administrators have increased their awareness about the costs and benefits involved with purchasing this type of equipment.

What is important for ergonomics teams in health care facilities to be aware of is not to get bogged down in the patient lifting issues. There are many other issues that can be addressed more easily, such as certain housekeeping tasks, the laundry, or trash handling.

Distribution centers — The primary task of manually picking orders can be very difficult to solve. This task has been studied thoroughly in many different situations and there does not yet seem to be a feasible way to resolve the core issues. You can modify racks to some degree, but at some point the order pickers need to bend, reach, and pick up a box in a way that is not ideal. Consequently, it is common for the ergonomics effort to be devoted to issues like the best rack locations for certain products, improving packaging design to make products easier to slide and lift (e.g., handholds), and various work method and administrative changes, like picking in layers and rotating pallets.

Delivery and on-site services — Those employers who provide services on-site to customers have a special problem in that the "workplace" is often someone else's and they have no control over equipment and conditions. Examples of employees who are affected in this situation include local delivery workers, couriers, repair personnel, and on-site customer service representatives. It can be awkward to evaluate these tasks as well as to suggest improvements.

Computer workstations — Working with computers is likely the most common task being performed today where there are widely recognized ergonomics issues. Fortunately, for the same reasons, the solutions are generally known and available, and there is not much need for brainstorming. Often the ergonomics process in these settings focuses on two items: (1) who needs what types of furniture and work aids, and (2) training on how to use the furniture and work aids properly.

Service industry — There is no difference between service industry tasks and manufacturing ones in this regard.

Trade associations

In many industries, the trade association is a natural sponsor for a great many of the ergo activities. An excellent example of how this process can be arranged is the American Meat Institute (AMI), the trade association for the meatpacking industry, which developed a comprehensive program in ergonomics to aid its members. Few, if any, other organizations can demonstrate an equivalent program involving as coordinated an effort on an industry-wide basis.[*]

Organization — Starting in the 1980s, The AMI Safety Committee, which is composed of member company safety directors, took on a strong and active role to coordinate activities. Furthermore, the AMI used its organizational structure to provide information and discuss issues with various company staff, such as the human resource managers of the member companies, the attorneys, the engineering managers, and most important of all, the Chief Executive Officers. The equipment suppliers who are associate members of the AMI became involved in the programs, thus providing a direct link between the packing companies and the vendors. Virtually all AMI conferences addressed ergonomics in one way or another.

Training — The AMI developed industry-specific training booklets and videotapes, including materials for both managers and employees. Thru the years, the trade association held numerous regional and national training conferences for safety committees and representatives of member companies.

Identifying improvements — The AMI took a variety of steps to promote problem-solving, including the following actions:

- Funded studies on knife design and optimal working postures through the University of Nebraska.
- Funded expert ergonomics assessments of meatpacking plants, shared with the industry.
- Met on several occasions with NIOSH scientists to discuss long-term research needs.
- Sponsored a meatpacking section meeting at the 1989 Human Factors Society Meeting.
- Described ergonomic innovations at the 1990 National Safety Congress.
- Funded a study on potential applications of new technology.
- Promoted cooperative R&D efforts among member firms.

These efforts paid off, as evidenced by the dramatic reductions in MSDs in the industry since the early 1990s.

---

[*] I was involved in these activities as the ergonomics consultant to the AMI thruout this process.

When You're Starting Out

Start slow and grow

It is important not to bite off more than you can chew. You may want to work with a small pilot area until you gain experience and establish a system to implement improvements. One of the worst things to happen is to claim a big project with a lot of fanfare, raise expectations, and then not be able to follow thru.

Sometimes it doesn't matter where you start. Just pick a project that you know you can fix, so that you can start building a track record of success.

Not a new program

Often it can be helpful to *not* present ergonomics as something new. To be sure, there are times when it *can* be helpful to institute the ergonomics process with a big flourish. The fanfare can attract attention and it signals that something new is coming. But this flashy approach can have a big downside.

Flavor of the month — The problem is that many companies have undergone a considerable amount of change in recent years and employees can be leery of yet another disruption. Worse yet are those organizations that have tried to make changes but failed. The new initiatives often begin with a lot of fanfare and big promises, but they fizzles out after a time. Even when the intentions are good and the outcome something desirable, it is easy for these failed attempts to be labeled "just another program." The more pejorative phrase is "flavor of the month."

Consequently, the experience in many organizations is that it is better *not* to promote ergonomics and other initiatives as something new. (This point should not be overstated. Many people welcome a change in physical working conditions and react by saying, "Finally!")

Good phrases — This predicament highlights the importance of communicating to everyone what you are doing. Specifically, the phrases you use in your explanations can be very important. There are ways to say things that signal there is a change coming, but one that is not disruptive and is more or less guaranteed to succeed. All of these phrases are based on the understanding that you have been doing "ergonomics" all your life, probably without realizing it.

- "We're going to formalize what we've been doing all along."
- "We're going to raise our efforts to a new plateau."
- "We have found that some initial attempts to make improvements have worked and we want to be more systematic."
- "Rather than just involve some people, we want to involve everyone."
- "This is the flip side of the quality process [or whatever]. Here's a good way to put into practice what we've learned.

Highlight past improvements

It can be very helpful to highlight past successes, whether you are starting the process from scratch or reinvigorating a process that has fallen dormant, or even trying to keep a good program moving along.

Past successes show a track record. They show you are not just making wild promises about what you would like to do, rather they document tangible successes. Moreover, the past examples help to demystify the field of ergonomics.

Thus, an early activity for you is to survey your facility for improvements that you have made in the past few years. Even if the changes were initiated for other reasons at the time, if they had an ergonomic impact, you can still include them if they help convey the message.

Simple and easy

Keep it simple. It was mentioned several times in the previous chapter that often the most effective activity is also the easiest. To reinforce the point, here are some examples:

- Often, the best way to identify problem areas and possible solutions is simply to talk to employees. You may not need any special surveys or studies. Plain conversations may lead you in the right direction.
- In my experience, the best task analysis tool is simply to watch videotapes of jobs and brainstorm improvements. You may not need much paperwork or any quantitative methods at all.
- Keeping a simple Ergonomics log is typically the best way of keeping track of issues, ideas, and progress. You don't necessarily need an elaborate filing or tracking system.
- The best solutions are often the ones that are simple and inexpensive.

Quick fixes

Quick fixes are important for three reasons: (1) to start establishing momentum, (2) to show that improvements do not necessarily need to be expensive, and (3) to help people to see their work areas in new light.

The best predictor of *future* success is *past* success, so if you are just starting, fix *something* and talk about that. It builds your credibility.

Demystifying ergonomics

In some ways *ergonomics* is a catchy term that appeals to our attraction for buzz words. But in other ways, it can be mystifying. There is almost always confusion about what ergonomics means. Two common preconceptions are:

- A mistaken view that ergonomics is a much narrower field than it really is, for example, something to do with chairs or computers, or synonymous with preventing MSDs
- A misperception that it a much more expensive and complicated field than it is, beyond the realm of ordinary people and ordinary companies.

Consequently, you may need to emphasize in the beginning that ergonomics is a practical problem-solving tool that makes considerable business sense if you work it correctly. It doesn't need to be hard or expensive, and it's not necessarily even anything new.

Pandora's Box?

There has been a concern in many companies that ergonomics programs might open a Pandora's Box, either by raising expectations too high for improvements in jobs or, more commonly, by resulting in massive reporting of injuries. In my experience, this has never happened. As described previously, most companies experience an increase in reports of MSDs, but this can be sign of success, and the numbers have never been anything close to a catastrophe.

But a more realistic fear is that employees will be skeptical and passive, rather than they will unleash a torrent of complaints and problems. You job will be to inspire, rather than restrain.

The biggest fears have been raised in companies with poor employee relations. The thing to keep in mind here is that a company in this situation can use ergonomics as a way to improve these relations.

The most common reaction of all, however, is that normally employees are rather realistic in their perceptions about what is possible. Communications is the key. You need to describe your plans in a way that grooms everyone's expectations.

Grooming expectations

There is indeed some balancing that you need to do. On the one hand you want to inspire people and raise their enthusiasm. But on the other, you don't want to raise anyone's expectations too high. Much depends upon your choice of words. You especially want to avoid promising anything that you can't deliver.

It is helpful to say things like, "I can't guarantee that I will implement your idea or fix your problem, but I will promise to listen to you and do what I can." Or, "We're going to start with baby steps so that we learn what ergonomics is really all about. Then we'll move on to bigger projects."

Sincerity

A couple of final points:

Be sincere. People can tell sincerity and it goes a long way in establishing a common bond that motivates without raising anticipations too high.

If you can, find ways to compliment your audience. You're there to help and that includes making people feel good about themselves.

Rules for Brainstorming

The importance of brainstorming has been highlighted thruout this book. The following materials can help you in case you need to familiarize people at your facility in the basic concepts.

Generate lots of ideas

It is important to generate a large *quantity* of ideas. Don't limit yourself or others. Say whatever comes into your mind and encourage others to do the same. Draw on your own experience, what ever you have seen or heard.

Combine, embellish, and build on the ideas of others. The more ideas that are shared the better. Whoever gets the *most number* of ideas wins.

Encourage harebrained ideas

Many people are fearful of saying something in public that would make them seem stupid. So the rules *require* you to say something off-the-wall. This is just a way of giving people permission to say something that they are not sure of.

Moreover, even though an idea may seem unrealistic or silly it has value. It may provoke thoughts from other people. A seemingly silly suggestion can spur an idea you didn't know you had!

Don't shoot down ideas

Don't focus on why someone else's ideas won't work. Your job is to figure out ways to overcome the roadblocks. There is plenty of time after the session to sort through all of the ideas and eliminate the ones that are not feasible. That's the easy part. The hard part is coming up with the ideas to begin with.

During the session you should not criticize any ideas because you may discourage others from contributing. Criticizing the ideas of others reduces the chance of finding the great ideas that make a difference.

Encourage everyone to participate

Everyone thinks and has ideas. Allow everyone to speak. Speaking in turn helps. Solicit ideas clockwise around the group. Encourage everyone to share his or her ideas.

Record all the ideas

Appoint a person to write down everything suggested. Don't edit the ideas. Write them down as they are mentioned. Keep a permanent record that can be used at future meetings. You may want to read through the list and take "inventory". This process sometimes stimulates more ideas.

Let ideas take time to grow

Once you have started brainstorming, ideas will come more easily. You are being creative. Give creativity time. Do not stop brainstorming sessions too soon. Let time go by and allow ideas to flow naturally.

This page available electronically. See Table of Contents.

Example of Brainstorming

Part I: What is wrong with this picture? Although the illustration of shoveling snow is "normal," the task requirements are not good. List as many issues as you can.

1. Awkward posture: severely bent back

2. Repetitive, forceful motions

3. High overall exertion

4. Poor environment: cold, wet, slippery

5. Poor job satisfaction (for most people)

In fact, many people suffer back injuries from overexertion while shoveling snow or from the severe compression forces placed on the spine while working in this posture. Furthermore, other people have experienced heart attacks from the high levels of exertion and metabolic load while shoveling. Shoveling snow can be a demanding task.

Part II: Study the illustration for a few minutes and write down every possible idea that comes to mind that might make this task easier. Several of the ideas must be harebrained answers in order to stimulate your thinking.

1. Use snowblower
2. Use push shovel
3. Move south
4. Have kids do it
5. Put canopy over steps
6. Use salt
7. Heat the steps
8. Use lighter scoop
9. Wait till Spring
10. Use broom
11. Use longer handle shovel
12. Build house level with ground
13. Use grating/let snow drop
14. Use pull-type shovel
15. Use flamethrower
16. Use leaf blower
17. Tramp down the snow
18. Use a ramp/tractor & plow
19. Provide good clothing
20. Use retractable steps

178

Best ideas

Once you have generated your list, you can find the one that is most appropriate for your situation. In this case, the best engineering solutions are probably putting a canopy overhead (which keeps the snow from reaching the steps) or heating the steps. If shoveling these steps were a task at work done eight hours per day, these engineering improvements would probably be cost effective. However, at home, that kind of expense would often not be feasible and a better shovel or snow blower would probably be a better way to go.

This message is one of the points of the exercise. Usually, there is no one "ergonomic" solution to a problem. Often, there are quite a number of ways to make improvements. The best choice depends on circumstances in each case. Moreover, you make that choice after the brainstorming is over.

Unconventional idea

One of the best ideas in this exercise is the two-handled shovel. This version cost about $15 and is actually just an extra handle that can be attached to any shovel, and indeed can be attached to a pitchfork in the summer for turning compost.

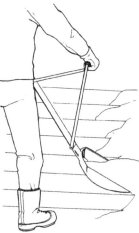

The two-handled shovel provides three important advantages. The extra handle:
• allows the shoveler to stand upright with the lower back in good posture and thus less risk of injury;
• provides considerable mechanical advantage to lift more snow and throw it further; and it
• provides more control over the shovel and allows improved ability to manipulate the scoop.

Thus, the two-handled shovel simultaneously improves (a) safety, (b) productivity and (c) product quality — quite a bargain for $15.

This is not to say it is perfect. Your forearm muscles can get sore and it doesn't make shoveling exactly fun. However, it's the shovel of choice for heavy, wet snow and for getting into the corners of the steps.

Not just another snow job

The real point here has nothing to do with shovels or snow. The point is how to come up with inexpensive ideas that work.

People have a hard time coming up with the idea of a two-handled snow shovel. What holds us back is our lack of imagination. We can think of ways to improve the shoveling of steps, but we have a difficult time thinking of adding a second handle on a snow shovel, since we have preconceived notions about what snow shovels are supposed to look like.

In our own workplaces, we also have a hard time coming up with alternatives. We all have preconceived ideas about how that work should be done, which inhibits our creativity. We become so used to seeing the routine that we cannot imagine the options.

To break through these barriers:

- Put on ergonomic glasses.
- Brainstorm, using a team process.
- Identify the ergonomic principle that addresses the problem, and then think through what kind of changes would be necessary to achieve that principle. (e.g., in this example, what does it take to permit the shoveler to straighten out the lower back?) The principles provide guidance in understanding how the task *ought* to look.

Training sessions

You can use this example, or any similar one, as a training exercise in your classes. If you present it right, it can get people into the habit of brainstorming and help reinforce the rules of brainstorming.

You can have all the participants start out on their own, and then after they have struggled for a while, turn it into a group exercise. Typically, the participants will identify many more ideas as a group than as individuals. There are, in fact, well over 50 different ways to improve the task of shoveling snow. Individuals usually can identify only a few ideas on their own. As a group, however, it is easy to come up with at least 20 ideas.

You can also turn this exercise into a contest. Divide your participants up into groups of three or four and explain that whoever gets the most ideas wins. You can emphasize that this is an exercise in *quantity*, not *quality*. The ideas don't have to necessarily be good ideas, just an idea.

Rules for Disagreements

Disagreements are inevitable

It's okay to have disagreements. In fact, it may be essential to have disagreements. Without disagreements you can get into a "groupthink" where everyone agrees, but does the wrong thing.

But disagreements can be positive or negative. Thus, you need to make sure your disagreements are constructive and not destructive.

Constructive disagreements serve a variety of good purposes. They can bring to the surface and clarify issues and goals and thus improve the quality of your problem-solving. They can act as a catalyst in your planning and can increase involvement of people. Often, when the group resolves the disagreements, both the organization and interpersonal relationships are better off.

Destructive disagreements on the other hand, divert energy from the real task and can destroy morale. Not only can they polarize individuals and groups, they can deepen the differences and obstruct improvements. They can even create suspicion and distrust and sometimes irresponsible behavior.

Rules

Consequently, we need to accept and even encourage disagreements. To help make sure they are of the constructive type, here are some rules:

Agree to disagree — The basic rule is that you can agree to disagree. This is a very helpful phrase that tends to remove emotions and personalities from the debate. It can keep anyone from drawing a line in the sand, and it signals that no one's psyche is being threatened, that both views are legitimate.

Compartmentalize — Part of the process is to learn to compartmentalize the debate. In a work setting (or maybe *any* setting), it is normal to disagree with a person on one matter, but agree with the same person a few minutes later on another matter. When you disagree with someone, you are not writing off that whole person; it's just a difference in opinion on one thing.

Remember that you could be wrong — Very few of us have never been wrong at least once in our lifetimes. It's OK to argue your point vigorously and with conviction. But don't get too carried away, since you could be wrong.

Understand the causes of disagreement — Finally, it can be helpful to dig beneath the surface of the disagreement. Are the antagonists arguing for different goals or different methods to reach the same goal? Similarly, does the disagreement represent a difference in style or a fundamental difference in values and objectives? Maybe the differences are in information and perception. Remember, too, that people play different roles, which is well and good. The union leader is *supposed* to worry about the effect on people. The senior manager is *supposed* to worry about money.

Immediate Action

It is not sufficient to have a great brainstorming session and then stop without establishing action items. Furthermore, you should constantly be seeking opportunities for action items that are immediate — something that you can do now, not next week. Here are two approaches to immediate action:

Trystorming

You *brainstorm* in the conference room, then you *trystorm* on the workplace floor.

Say that TeamErgo just thought of a great idea for an unconventional device. A good immediate step is to build a prototype to test the idea. You use cardboard, tape, unused pallets, and anything else that's handy to rig up a model. That's *trystorming*.

Examples of successful trystorming include:

- A distribution center that used sheets of plywood to test the idea of sliding heavy products rather than having to pick them up and carry them.
- A packaging operation that used cardboard to build a slanted stand in order to try sliding the products into shipping boxes.
- An assembly plant that mocked up a new workstation with cardboard.
- A manufacturing plant that used boards taped to a swivel chair to test the idea of using a turntable.

In all of these cases, the mock-ups were dismantled after the test. However, they proved that the idea would work and made it impossible for the naysayers to spread their pessimism.

Just do it

At other times, TeamErgo may identify a very logical idea that everyone agrees is a good one. Rather than wait, it may be appropriate to immediately go ahead and make the changes. Examples of these types of instances are:

- A department where a couple of ErgoTeam members, the supervisor, and an employee changed a poorly laid out work area into a rational system in about two hours.
- A supervisor, after realizing that a wooden table was too high and was causing problems, immediately fetched a power saw and cut down the legs.
- At a manufacturing plant, team members went into the plant junkyard, found a section of conveyor that was the perfect solution to a work area, carried it into the facility, and started using it.

Obviously, you want to make sure you have qualified people involved and that you follow normal organizational rules. The point is to strike while the iron is hot.

More Guidance on Problem-Solving

As you proceed in the process, it can be helpful to keep
in mind a number of points:

It's hard to be perfect

It is extremely difficult to design a job to be ergonomically perfect. It can be exorbitantly expensive to get everything exactly right, plus there can be conflicting issues that are difficult to resolve. Thus, it is important not to get bogged down trying to fix everything all at once.

- Remember the 80:20 rule — Often, a few changes can fix 80% of the problems. It can take a lot more effort to fix the final 20%.
- The concept of continuous improvement is also important. Do what you can, even if it means making a small change now and then a larger change later. This is all a normal part of engineering and design.

Ergonomics in advertising

Beware the advertisement that says "ergonomically correct." You must understand the underlying principles of ergonomics to be able to evaluate products.

- "Ergonomic" products that are used inappropriately are not ergonomic.
- There is no certification or testing procedure to determine if a product is truly ergonomic . . . buyer beware!
- *Any* product or change that helps adapt a workstation to fit individual needs can be "ergonomic."

Focus on the fixable

You may not necessarily want to start with the hardest projects. You may want to focus on fixable things instead.

A good example is in hospitals and nursing homes. The biggest cause of injuries is lifting patients. But this is also a very difficult problem to fix and it is easy to get overwhelmed. You may be better off, at least as a start, by looking at the laundry or housekeeping issues, since they tend to be easier to resolve. Once you have a track record in solving these other problems, you may be in a better position to focus on the harder items.

Another way of saying of saying the same thing is: After you conduct an analysis of the worst problems (i.e., create a "Pareto chart"), skip over the first couple of items and go down to the third or forth ones on the list. These are still important issues, but chances are, they will be easier to fix.

A hundred pair of eyes

A clearly successful strategy has been not to rely solely on experts to identify improvements. Involving people in the effort to evaluate tasks and brainstorm improvements has led to thousands of good ideas.

There's an "ergonomist" in every one of us. Almost everyone has an idea about how a job could be made better.

Improving repetitive jobs

The term *repetitive motion* is somewhat of a misnomer and can be very misleading. First of all, there are other factors such as heavy exertion or static loading of muscles that can be more important problems than anything repetitive. Secondly, trying to eliminate the repetitive nature of many jobs is extremely difficult and you can easily be stymied by attempting to do so. It is much easier to think about eliminating individual motions or changing from awkward and harmful motions to smoother, easier motions.

A third important point is that even if a problem is caused by the repetitive nature of the task (that is, the task is not a problem at all when done only sporadically), it does not necessarily follow that you need to eliminate the repetition to solve the problem. Rather, the solution more commonly is found by reducing the extent of awkward postures or heavy forces.

So when studying jobs, focus on *motions* as well as factors such as static load, high force, pressure points. The term *repetition* is not helpful at this point and may lead you astray.

Don't fix anything that you shouldn't be doing in the first place

It is not unusual for some tasks to be performed even tho they are totally unnecessary. Usually this situation occurs because a product has changed and a step that used to be needed no longer is, but no one has realized it. Or a new process or piece of equipment has been added elsewhere in the facility that could be used to perform this step much more easily. A good task analysis should uncover these instances.

Fresh pair of eyes

One of the great advantages of the team approach is that you can involve a fresh set of eyes. Typically, when you evaluate a job using a team, you include someone who is very familiar with the operation and someone who isn't. The people who are not familiar with the operation are often able to ask simple, yet surprisingly penetrating questions that can get people to think. For example, just asking, "Why do you do it that way?" can challenge some assumptions that might not be true.

An atmosphere of creativity

In many facilities, the best solutions (the low-cost ideas that simultaneously increase efficiency while improving employee well being) are often not obvious in the beginning. Thus, as part of a successful ergonomics program, it is important to create an atmosphere of creativity:

- Willingness to question past methods
- Enthusiasm in considering new ideas
- Encouragement of harebrained ideas

Nothing beats videotaping a job, then sitting in a conference room with a group of employees, supervisors, and engineers to brainstorm possible improvements. The brainstorming process and a willingness to proceed with trial and error should become part of normal businesses.

Resurrect past efforts

It is not uncommon during brainstorming sessions for some-one to say "we tried that a few years ago, but it didn't work." But, typically, no one can remember *why* it didn't work. Consequently, it is worth investigating again. You need to determine if it was a bad idea or if it was merely that the implementation process failed at the time.

There can be any number of reasons why something didn't work in the past other than the idea wasn't good. It could be that the people trying to implement the change didn't follow thru or got pulled away to work on something else. Or it could be that no one explained to the employees why the change was being attempted.

Another common reason why things might not have worked in the past is that whoever was in charge simply lacked the energy to continue. Trying to make a new concept work can take a lot of energy and it is rather normal to get tired of trying and move on to something else. In this case, a new team with fresh enthusiasm may be able to resurrect the past idea and make it happen.

Never assume

It is common for people to assume that a particular task is already set up in the best possible way. It is easy to take things for granted, such as by assuming that someone previously studied all the heights, reaches, layout, tools, etc. and designed the optimal arrangement. Even if there are obvious problems, it is all too easy to wrongly assume that someone has already scientifically determined that there are no better alternatives.

- **Operations change** — A particular setup could well have been needed a decade or so ago, but with changes in the operation (new products, upgraded technology, etc.) that need may have disappeared. It might be possible now to make an improvement that was originally impossible.

- **Facilities change** — It is not uncommon for particular operations to be shifted from one facility to another (or one part of a facility to another part). Typically, the setup in the new site is made in exactly the same way as in the old site. However, it is also typical that the space available is dramatically changed. Again, this can provide opportunities to improve the setup in ways that were not possible beforehand.

- **We learn** — Sometimes a task was set up in what was then thought to be the best possible way. But now, with a new set of tools — your ergonomics glasses — you can see ways of making changes.

- **We forget** — It is surprising how often work areas were set up to be temporary and thus with known problems. But people get distracted, then forget, and then become so used to seeing the area, that the problems become "invisible."

It's Worth Doing the Little Things

The multiplying effect

The graph below indicates that there is a multiplying effect between exertion and repetitive motions on hand/wrist MSDs.[*] Forceful motions alone and repetitive motions alone can each increase the risk for an MSD by about three times. But jobs that involve both forceful and repetitive motions increase the risk by almost 17 times.

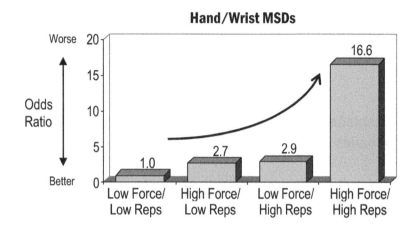

Hand/Wrist MSDs

The dividing effect

The same graph also makes a different point, one that is much more optimistic: There is a dividing effect; you can go the other way.

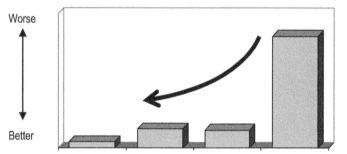

Big impact

Since the combined effect between the different risk factors seems to be on an exponential curve, there are a number critical implications:

- If you can affect just one of the risk factors, you can have a big impact. You don't need to make jobs perfect.
- A little bit can go a long way — it's worth doing the little things.

This graph supports my own experience that in many workplaces it doesn't take much to have a major influence. It makes the job of being involved on TeamErgo very satisfying, since you can a difference in the well-being of your fellow employees.

---

[*] Silverstein, B.A., *The Prevalence of Upper Extremity Disorders in Industry.* Ann Arbor: Center for Ergonomics, University of Michigan,1985.

Stop and study

Much of the workplace ergonomics process is not technically difficult. Most of the time you don't need to know formulas or complex issues of human physiology. The difficult part is that people get so in the habit of doing a certain task or seeing it being done, that we don't "see" it at all. We don't notice or challenge the things that we are accustomed to. Consequently, we must force ourselves to take a step back from whatever we're doing and *think*.

Quick fixes

Always be on the lookout for quick fixes. A lot of small improvements can add up and make a difference in the workplace. Perhaps more importantly, the quick fixes can get people in the habit of thinking and suggesting changes.

Old-fashioned ergonomics

Don't forget that you have probably been making ergonomic job improvements for years — you just never used the term. Examples include moving things around to reduce the amount of reaching or using power tools to reduce repetitive motions or high exertion.

Ergonomics in everything we do

While the focus of these materials is the workplace, off-the-job issues are important, too. It is good to get in the habit of seeing *everything* with your "ergonomics glasses" — home chores, hobbies, and leisure activities.

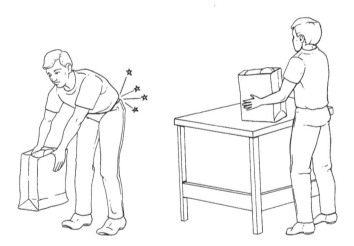

Why work the hard way? Rather than putting things on the floor where you will just need to pick them up again, why not put them on a table, a shelf, or a chair?

Chapter 7
More Guidance on Specific Issues

In the two previous chapters, we have reviewed the basic elements of an ergonomics program and then moved on to describe the workplace ergonomics process. This chapter provides more background detail on a number of topics:

Job rotation — Job rotation may seem like an easy way to reduce the risk of MSDs, but setting up a meaningful system can in fact take a lot work. If you want to start rotation, this section provides some guidance.

Be aware that altho job rotation can work well in some settings, there are many shortcomings and pitfalls to this approach. Job rotation alone does not change the risk factors present in a facility; it only distributes the risk factors more evenly across a larger group of people. Moreover, it does little by itself to stimulate innovation and efficiency.

Back belts — Another seductive approach with pitfalls is the use of back belts. There is some indication that they work in some settings, but most of the evidence is negative. Above all they do nothing about the source of the problem or to stimulate innovation and efficiency

Setting priorities — This section provides some guidance and background information on how to formally go about setting priorities.

Analyzing MSD records — Similarly, this section describes how to analyze MSD cases at your workplace to identify any patterns.

Expect increases in MSDs — In the initial phase of a program, you increase awareness among employees, which sometimes leads to an increase in reports of MSDs. This can be good, if you prepare for it.

Reward systems — Sometimes there are conflicts between the way that bonuses are offered for low numbers of injuries and the benefits of early reporting of MSDs. Here are some comments and suggestions for resolution.

Ergo log — A sample log is included here, along with suggestions for its use.

Videotaping jobs — This page provides some tips on using the video camera.

Ergonomics consultants — If you are considering the use of a consultant, here are tips on getting the most from the relationship.

Written program — The chapter ends with tips on writing a formal plan and samples of different ways to write a plan.

About Job Rotation

Introduction

There are many reasons for implementing a job rotation system, including the potential for increased flexibility in production, increased employee satisfaction and lower MSD rates. However, establishing a rotation system that properly determines job rotations and monitors their safe use is not a simple task. There are many issues to consider and no official protocol or methodology to call upon. The successful implementation of a program requires teamwork from all parts of the organization, including management, union, medical providers, and especially the employees themselves.

Many job rotation systems have failed because of lack of planning and lack of foresight into the problems and shortcomings of rotation. It can prove more difficult than it might seem at first glance, since it involves changing the organizational structure of an entire facility.

The following materials provide systematic guidance for setting up a site-wide rotation system. This guidance should be viewed as a starting point for further discussion by workplace personnel.

Roadblocks

There are two major categories of roadblocks that are often encountered in setting up a job rotation system:

Cultural issues — the first set of difficulties are associated with the challenge of changing the work structure and not from the job rotation in and of itself. Examples of problems include:

- Experienced workers not wanting to learn new types of work
- Employees not wanting to "lend" their equipment to others
- Pre-existing differences in wage levels among employees whose jobs are to be rotated
- High-seniority employees who have "paid their dues" working at difficult jobs may believe that they have earned their right to easier jobs and may resist going back to more difficult work.
- Practical problems of physically getting from one job to the next

Rotation issues — The other set of difficulties have to do with issues surrounding the rotation schedule itself:

- Difficulties in finding appropriate jobs to rotate to (for the goal of reducing MSDs)
- Difficulties for employees in learning the subtleties of some tasks and thus end up *increasing* the physical demands.
- Inability of some employees to be physically able to perform the most difficult tasks
- Education and training of workers for new jobs
- Inconsistency of application

Basic limitation

Job rotation alone does not change the risk factors present in a facility. It only distributes the risk factors more evenly across a larger group of people. Thus, the risk for some individuals can be reduced, while the risk for others can be increased. However, there will be no net change in risk factors present. This can be shown in the following graph.

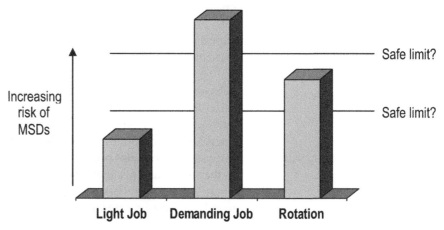

When employees rotate between two jobs the risk exposure can be thought of as being "averaged." Job rotation may drop the average to within a safe level, or raise the whole group in excess of safe limits. Unfortunately, it is not possible with current knowledge to determine what the safe limit is. For this reason it is prudent to be cautious about job rotation. Engineering changes should remain the goal of the ergonomics program.

More limitations

If the jobs being rotated involve the same muscle-tendon groups then the benefit of MSD risk reduction is lost. Thus, rotation among jobs that are similar is not appropriate. Situations that are best able to benefit from job rotation are those where, for example, a wrist intensive task is adjacent to a back-intensive task.

Additionally, if the rotation is too infrequent, such as a daily rotation, the benefit is also lost. Typically, employees should rotate every two hours. An hourly rotation is probably better and a four-hour rotation probably the maximum that would provide any benefit from an MSD perspective.

OSHA guidelines

The following is excerpted from the OSHA *Ergonomics Program Management Guidelines for Meatpacking Plants*:

> Job rotation should be used with caution and as a preventive measure, not as a response to symptoms. The principle of job rotation is to alleviate physical fatigue and stress of a particular set of muscles and tendons by rotating employees among other jobs that use different muscle-tendon groups. If rotation is utilized, the job analyses must

be reviewed by a qualified person to ensure that the same muscle-tendon groups are not used.

A "qualified person" is one who has thorough training and experience sufficient to identify ergonomic hazards in the workplace and recommend an effective means of correction; for example, a plant engineer fully trained in ergonomics - not necessarily an ergonomist. In analyzing jobs for rotation, the qualified person must have sufficient expertise to identify the ergonomic stresses each job presents and which muscles and tendons are used.

Job rotation can mean that a worker performs two or more different tasks in different parts of the day (i.e., switching between task "A" and task "B" at 2-hour or 4-hour intervals). The important consideration is to ensure that the different tasks do not present the same ergonomic stressors to the same parts of the body (muscle-tendon groups). There is no single work-rest regimen that OSHA recommends; it must be determined by the nature of the task.

These excerpts indicate the importance of establishing a formal, documented job rotation system which carefully matches jobs. This matching system should ensure that different muscle-tendon groups are involved.

Scoring system

For best results, it is important to quantify or score the risk factors associated with each of the tasks that are to be rotated. There is no established system or protocol for these scores and you will need to select or develop a system that is appropriate for your site and the tasks in question.

Typically, a score would be calculated for each job for (1) the hand and wrist, (2) the arm and shoulder, (3) the lower back, and (4) the overall job difficulty. However, other factors and body parts may need to be taken into consideration depending upon the tasks.

Whatever scoring system is used, it can be helpful to convert your final results into "red," "yellow," and "green" to represent high, medium, and low risk. Thus, a good rotation would a job with a red score for the lower back and one with a green score for the lower back.

Be systematic

To realize the beneficial aspects of job rotation it is necessary to establish definitive internal guidelines that insure consistent application and at the same time allow for restricting employees from rotating into jobs they cannot perform. To ensure that all job rotations meet basic ergonomics requirements a consistent and systematic approach is required.

It is probably best to start slowly at first, such as in a pilot work area so that the program can be further refined before being implemented elsewhere.

Steps for Implementation

Step 1: Hold an employee meeting to determine interest and gain involvement and input. During this meeting it would be appropriate to have a short presentation on ergonomics and job rotation. The purpose here is to build upon the ergonomics training already received and further it by discussing the relationship between it and job rotation. At this time it would be appropriate to issue a Job Rotation Questionnaire (see following pages).

Step 2: Calculate the scores for the jobs considered for rotation. Use these scores to establish which jobs should be rotated with which. In general, decisions about the suitability of a particular job rotation should be based on the following:

Job A Rank	&	Job B Rank	=	Decision
Red	&	Red	=	Unacceptable
Red	&	Yellow	=	Unacceptable
Red	&	Green	=	Acceptable
Yellow	&	Yellow	=	Acceptable*
Yellow	&	Green	=	Acceptable
Green	&	Green	=	Acceptable

not recommended; try to avoid if possible

Step 3: — Apply a common-sense review to ensure that the logistics of the proposed rotation are suitable and that the job rotation seems reasonable. Also, review the job rotation scheme with the affected employees. The employee concerns and insights should be taken into account. If necessary, changes to the list should be made, and final approval for the list obtained.

Step 4: — Provide employees with any training that they may need to perform the tasks or handle the tools and equipment. In general, experienced employees going to a new job should receive the same training requirements and documentation that a new hire must have before starting in that position.

Step 5: — Provide employees with adequate break-in time to ensure that they are fully qualified and physically conditioned to perform their new tasks. Similar to training requirements, the same guidelines for new hires starting out should apply to experienced employees starting in a new job. Even if the employees have performed the job previously, they should generally be allowed the break-in period to become accustomed to the work again.

The training and break-in period enables the employee to develop those subtle work techniques needed to perform the task the easiest way and thus minimize the risk factors. This also suggests that the number of jobs included in a particular rotation should be kept to a minimum, perhaps two or three, allowing the employees to become "experts" at each task.

Step 6: — Begin job rotation.

Step 7: — Monitor the new rotation to ensure flexibility and consideration for individuals that are having difficulty performing new tasks. Assess if further training, break-in, and/or accommodations can be made for these individuals.

Step 8: — Hold follow-up meetings with employees to evaluate the job rotation. Survey the employees using the job rotation questionnaire again. Compare results to the initial survey. If results are favorable then continue rotation. If results indicate a problem then decide if corrective action is needed or if rotation should be discontinued.

Step 9: — Track other measures such as injury rates, turnover, employee satisfaction, or workers compensation to determine effects of the job rotation.

Comments

These steps should be viewed as options and starting points for further discussion by the site ergonomics team and other interested personnel. This framework was written with a particular company and industry (meatpacking) in mind, and you may have different needs and applications. The objective here is to show you one approach for developing a formal, consistent, and systematic method of job rotations that are based on the requirements of the jobs.

To help you make sure that all of the steps of the process are completed and documented, you may find it helpful to use the *Job Rotation Checklist* found on the following page.

Role of TeamErgo

Anyone should be able to suggest job rotations, including supervisors, production employees, or union officials. However, the job rotation scheme should be approved by TeamErgo with input from the affected employees before being implemented.

Job Rotation Checklist

The following must occur for each job rotation setup.

Jobs proposed to be rotated are:

_____

_____

_____

_____

___ Has an employee meeting been held to determine interest and gain involvement and input?

___ Has each task involved in the proposed rotation been scored to determine precise requirements?

__ Has common sense been used in evaluating job rotation possibilities?

__ Have all employees involved in the rotation schedule been trained to do all tasks?

__ Have all employees been provided an adequate break-in period to insure they are:

    __ fully qualified to do all tasks?
    __ physically conditioned and accustomed to do the job?

__ Has flexibility and consideration been given for individuals in the rotation schedule? Are there any employees who would have physical difficulty in performing all the tasks? Can accommodations be made for these individuals?

__ Have formal follow-up evaluations using TeamErgo members and supervisors been conducted?

__ Are benefits or problems being tracked (increased or decreased injury rates, turnover, employee satisfaction, workers compensation costs, etc.)?

Keep the above documentation on permanent file.

Job Rotation Questionnaire

Name: _____ Date: _____

Department: _____ Job Title: _____

1. Are you currently rotating jobs?   Yes   No
   If no, go to the next question.
   If yes:

   a. Do you like it?   Yes   No

   b. If no, why not?
      _____
      _____

   c. To what jobs do you rotate?
      _____
      _____
      _____

   d. How often do you rotate?   1/2 hour   hourly   two hours   other _____

   e. Have you received appropriate training for the jobs that you rotate to?   Yes   No

2. If you answered No to question 1:

   a. Would you like to rotate?   Yes   No

   b. If No, why not?
      _____
      _____

   c. If yes, to what jobs would you like to rotate?
      _____
      _____
      _____

4. If you indicated on the form that you were having discomfort, have you seen the nurse?

5. Are there any other comments that you would like to make?
   _____

About Back Belts

Theory

When you lift, your stomach muscles tighten and form a sort of cylinder around your spinal column. This cylinder stabilizes the spinal column and tends to offload some of the weight pressing down on the discs. The theory behind the back belt is that it mimics and supports these abdominal muscles. Normally, you would wear the back belt loosely, and then cinch it up when you are going to make a lift.

Studies

The studies are a bit mixed, but mostly negative about the effect of back belts. Some internal company data show a positive effect in reducing injuries, but these companies also instituted a number of other activities like training and changing administrative practices at the same time. Consequently, it isn't really clear what specific action(s) lowered the injury rate.

The National Institute for Occupational Safety and Health (NIOSH) has reviewed the literature and conducted studies of its own and has concluded that there is "insufficient evidence to recommend the use of back belts as a back injury prevention measure." Some workplace safety experts are also concerned that back belts might actually do harm, in that they may inadvertently weaken stomach muscles or create a false sense of security among employees.

Reactions

Professional ergonomists (including me) tend to be a bit biased against back belts because they do nothing about the source of problems. Since their purpose is to shore up the employee to perform tasks that exceed human capabilities, back belts are the antithesis of ergonomics.

On the other hand, there is often good acceptance among employees for wearing back belts and they may serve a good purpose of reminding employees to lift using good technique. That is, when you bend over to lift, the back belt binds your torso and you are reminded to lift correctly.

Giving up

The biggest problem, however, is that when you rely solely on back belts as the solution to injuries, you give up all hope of discovering a better way of doing a job. As long as you are studying the tasks, brainstorming, and trying out ideas, you have the opportunity of increasing efficiency *and* reducing injuries.

As an example, ditch diggers 100 years ago must have had very high rates of back injuries. Those injuries were almost completely eliminated, not by using back belts, but rather by introducing ergonomic improvements like backhoes and trenching machines that have made construction work incredibly efficient. Ideally, in the world of ergonomics, equivalent mechanical innovations will accomplish the same in other industries.

Tips on Setting Priorities

Quick fixes

Whenever possible, go ahead and make changes that are relatively easy and inexpensive. But the more difficult ones may take some deliberation.

Severity first

It is typically not realistic to address all jobs at once. You must make a decision on where to focus your energies.

Sometimes, one area clearly emerges as a good target area based on your own knowledge or discussions with site personnel, such as supervisors or employees. In these cases, everyone more or less knows where the problems are and there is no need for further study.

At other times, you might need to use a more formal approach. Setting priorities usually means taking the most severe problems first. Unfortunately, severity can be defined in a number of ways, including situations where (1) a *few* people are affected by *extreme* problems, and (2) the opposite, where *many* people are affected by *moderate* problems. Both are valid approaches, but your choice of targets may end up being different.

To complicate matters, "severity" can be gauged on problems identified via any of your background assessments depending upon the circumstances:

- injury/illness records
- workers compensation costs
- employee surveys
- walkthrough surveys

Feasibility, too

Furthermore, it is often crucial to take into account the feasibility of making improvements when selecting target work areas. You may choose to focus on less severe problems that promise easier fixes. Often times, the worst problems are that way for a reason — they are difficult to solve.

Especially if you are just starting out, you generally will want to pick projects that will lead to easy victories so that you establish a track record, gain confidence, and build credibility.

Making informed choices

There is no standard mathematical formula for comparing various measures to set your priorities. At some point, you will probably need to make a judgment call.

These are tools to help you gain insights about your operations and make informed judgments. Results do not necessarily need to be followed rigidly once compiled.

Nonetheless, there are some techniques that can help you make informed choices. The following are two such methods.

Ranking

One approach is to compare the summary results of each of the background assessments. Often, a computer spreadsheet works well to help rank work areas. To do this you can:

1. List the criteria you are taking into account in making your decision across the top row.

2. List the departments (or jobs, lines, etc.) in the left-hand column.

3. Enter the appropriate numbers from your background assessment.

4. Sort each column in rank order, printing the ordering of departments for each sort.

5. Compare the rankings and printouts.

The spreadsheet would look something like this (you might choose different factors to include):

Dept.	Cases	Rates	Survey Score	etc.
A	32	2.3	8	
B	5	4.1	12	
C	12	2.8	6	
etc.				

Once your data is entered, you would use the "sort" feature in the spreadsheet to rank each category. Then you print out each of these rankings and look at how the jobs are ranked depending upon which criteria you used.

Occasionally, the same department tops each list, no matter which criterion is used for ranking. This would be a department that affects a lot of people and has extreme problems. In this case, your priority is clear.

More often, the areas that top each list differ somewhat depending upon which criterion is used. Then, you must make an informed choice.

There are pros and cons for each method of ranking and there is no "correct" choice. Each view provides its own perspective and helps you understand more about your workplace. Just make sure you recognize the rationale for setting your priorities and remember that there may be other areas that need attention after you have finished improving your first target area.

Scoring

An additional way to formally identify priorities is to score each category on a scale of 1 to 3 (or 1 to 5 or whatever range you wish). Again, a number of different criteria can be chosen.

Category	Description	Score
Number of employees affected	Many	1
	Handful	2
	One employee	3
MSD cases	>20	1
	5-20	2
	<5	3
MSD lost, restricted, & transfer days	>50	1
	10-50	2
	<10	3
MSD rate	>75	1
	50-75	2
	<50	3
Feasibility of improvements	Easy	1
	Moderately hard	2
	Long term	3

Note that all of the numbers in the example above are arbitrary. You should select your own based on the range that you are experiencing in any particular category.

Also, note that the descriptions are made in such a way that number one is always the highest priority. The scores can then be placed in the spreadsheet, then added (or multiplied) across the rows. The final column can then be sorted to gain an overall ranking of operations. In the following example, Operation C ends up having the highest priority.

Opera-tion	# of employees	Cases	Lost Days	Rates	Feasibility Rating	SUMMARY SCORE
A	2	2	1	1	1	**4**
B	3	3	2	2	3	**13**
C	1	2	1	1	1	**2**
etc.						

Scores ⌐

Add (or multiply) the scores across each row, then sort by the last column.

Based on the above exercises, you can select target areas for focused problem solving. This will lead to further evaluation and additional decisions — such as cost/benefit analysis of various options for improvement.

To reemphasize, when using either of the above methods, you will be making judgments, and no one approach is the most correct and scientific. These are tools to help you gain insights about your operations and make informed decisions.

Pros and Cons of Measures

Criterion	Pro	Con
Number of OSHA-recordable cases	Easy number to obtain	(1) Tends to make large departments look worse and small departments better (2) If there is an active medical program, the number of recordable cases may rise simply because of higher awareness among employees; consequently these numbers may not indicate the severity of the problem at all
Incidence rates	Enables comparisons between small and large departments	Can exaggerate small groups
Lost days	Strong indicator of relative severity	A few extreme cases can create a false impression
DART (Days Away from work, Restricted work activity, or job Transfer)	OSHA's new method to account for severity; probably now the best single measure to use; can be tallied by cases, days, or rate	Still doesn't account for everything, so other measures are nonetheless helpful
Employee survey results	Theoretically, these are more sensitive than reported cases of injury/illness	Subjective and based on perceptions rather than actual consequences
Walkthru survey score	Based on direct observations of job conditions	Without elaborate studies can ignore relative severity
Active medical surveillance	Probably the most accurate	Difficult to do and hardly ever done (but may increase in the future)

How to Analyze Your MSD Cases

Background

For accurate use of injury statistics in the ergonomics process, you need to separate out the MSDs from the full list of injuries and illnesses. The primary reason for this is that the factors involved in their causation are different from those related to other injuries and illnesses. Consequently, the overall injury/illness rate may not reflect the trends for MSDs.

Unfortunately, you must do much of this work by hand, since most computer-based searches can be inaccurate (most notably because a query for "back strains" usually provides injuries from slips and falls in addition to lifting or pulling). On the positive side, however, doing the work by hand involves reviewing each case separately, which can lead to better understanding of the injuries.

Results from this type of analysis can help you in a number of ways, including helping you to:

- establish priorities for job improvements.
- compare jobs, departments, or facilities.
- track progress.

Remember that one of the main reasons that you keep a record of injuries is to develop a data base. And data bases exist for the purpose of analyzing to see if there are any patterns.

Simple version

If you work at a small company or just want to make a quick study of your MSDs, all you need is the Injury and Illness Log. In the U.S., this log is called OSHA Form 300 and is the basic government-mandated form for work-related injuries.

Review roughly a year's worth of injury records, generally the more the better.

Identify those cases that can be considered in the category of MSDs. A nurse or someone familiar with medical terms may need to be involved. Include:

- Classical repetitive motion disorders — tendonitis, carpal tunnel syndrome, etc.
- Strains and sprains related to lifting, pushing, and pulling. These are usually back strains, but also include shoulder, wrist, and thumb/finger strains related to heavy lifting, pushing, or pulling.
- Hernias, abdominal strains and similar disorders that are directly related to an ergonomics issue like heavy lifting, pushing, or pulling.

Do not include:
- Strains and sprains that are related to slips, trips, and falls
- Acute (instantaneous) injuries, like bruises or cuts
- Noise-induced hearing loss or chemical exposures

Be aware that sometimes the forms might only contain a vague statement about "sore wrists" or "backache" rather than a formal diagnosis. For the purposes here, this description is sufficient and you should include it.

Write down the job title, department, shift, the type of MSD, the number of lost or restricted days, if any, and any other relevant items. Include the number of lost days or restricted days, if any. Your list should look something like this:

Job Title	Location	MSD Description	Days Lost	Days Restricted
Assembler	Dept 12	epicondylitis	4	–
Palletizer	Pack	back	8	4
Packager	Pack	sore wrist	–	2
Assembler	Dept 10	back strain	3	–
Assembler	Dept 10	sore wrist	–	1
Assembler	Dept 12	bilateral CTS	15	14
Packager	Pack	pain in wrist	–	–

Tally your results. Which job title is most affected? Which location? What type of MSD is the most common? It is often helpful to use the table format shown on the page after next. Note that you may need to make a separate table for "recordable cases," "DART," or any other similar category you used.

Use results to set priorities for improvement or to make various comparisons.

More complex version

If you want to be more accurate or determine the costs associated with your MSD cases, there are additional steps that you can take. Typically, there are several forms that you need, depending on the extent of your study.

- **Individual reports of injuries** — This is OSHA Form 301 or its equivalent (since many companies have their own forms for the same information). Sometimes it is called First Report of Injury or may have other names. These reports are important because they contain a narrative of the injury, which helps you determine if the injury is related to an ergonomics problem or not. That is, the description of, for example, a back strain should say whether it was from lifting a heavy load (and thus include) or from slipping in the parking lot (do not include). Sometimes there is a great wealth of additional information in these reports.

- **Incident log** — Many companies keep a record of *everything* that is reported, including "near misses" (an accident where no one was hurt, but easily could have been) and reports of aches and pains that are not significant enough to meet the criteria to be recorded on the OSHA 300 Log. Sometimes it is called the Nurse's Log or even the Accident Log.

- **Workers' compensation annual report** — The actual name of these records or report varies considerably in different companies, but usually it is available in your Human Resources Department or directly from your insurance carrier. You need the list of individuals, by name, and the costs associated with each. Typically, this report contains both an annual cost and a multiyear cumulative cost, as applicable.

Approach

Past data — To do a study of past trends, you need to find the reports from previous years. You should get roughly 100 MSD cases for an analysis, the more the better. In smaller facilities, this may mean getting two or three year's worth of records to create a sufficiently large data base to get meaningful results.

Future data — Once you catch up to the current date, it is easier to review the reports as they come in. For example, you could request copies of all injury reports each week.

Your database — You should establish your own spreadsheet for this purpose. This spreadsheet should look like the list shown in the "simple version" above, but would include additional columns for more information.

Steps

Read the narratives from the Individual Reports of Injuries and review the Incident Log. As described above with the "simple version," identify those cases that appear to be MSDs.

Enter the data in your spreadsheet. Include as many categories as you see fit. For example, if your study is for multiple years, obviously you need to include a category for year.

Sometimes as you read thru the injury report narratives, you start noticing special patterns, like a particular tool that seems to be causing problems. You may want to make up a new category for this type of information and enter the data in your spreadsheet.

If you want to include workers' compensation costs, you will probably need to include the employee name, date of report, accident report number, or other identifier so that you can keep your data straight. (Normally, you would not generate any reports that have individual names listed, but you may need to include names in your data base for clarity.) Once you are done generating this list of individuals who reported MSDs, you turn to the workers' compensation reports, look up those individuals, and note the costs associated with those cases.

Be forewarned that typically the workers' compensation records are in a completely different format than the injury records and your study can become very frustrating at this point. For example, the injury records might be kept by calendar year and the workers' compensation costs by fiscal year.

MSD categories

Often it is helpful to lump the MSDs into three or four types, otherwise there can be too many names to yield meaningful results. For example, all back strains, hernias, and abdominal strains might be lumped together as "back/torso" or simply as "backs." This is an accurate lumping for this purpose, since all of these disorders would typically be related to manhandling a heavy load. Similarly, you can consolidate carpal tunnel syndrome, "sore wrists," and anything else related to the hand and wrist under a single term.

Analysis

Now you are prepared to make comparisons in a number of different ways. Use the spreadsheet sort feature for each of your categories.

- Total number of cases on the incident log, OSHA log, and workers' compensation lists, respectively.
- Total DART cases, days, and rate
- MSDs by location, shift, etc.
- Workers' compensation costs for MSDs
- All of these can be reviewed by year.
- It can be helpful to take into account the number of employees in the various comparisons so that you can calculate *rates*.

Start generating tables for the totals of each sort that you make. Depending on the amount of data you have and the number of categories you set up, you can generate dozens of tables like the following:

MSDs — DART Days — Two-year period

Job	Wrists	Arm	Backs	**Total**
Job A	85	12	115	212
Job B	11	–	–	11
Job C	–	48	39	87
Total	96	60	154	**310**

A two-year review of MSD lost days shows that Job A has the greatest risk of injury, affected by back injuries and wrist problems. Job B appears to have a wrist issue and Job C a problem with arms and backs.

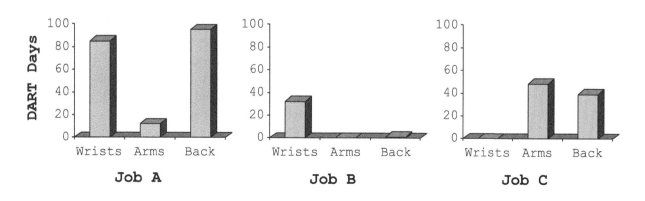

Draw graphs

As shown in the example on the previous page, trends and differences are much easier to notice on graphs than by comparing numbers alone. Get in the habit of generating graphs.

Make sure when you make multiple graphs to compare different items that you use the same scale on the vertical axis. Otherwise, the graphs may be visually misleading.

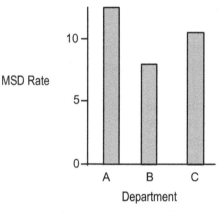

Bar graphs are good for comparing departments or facilities.

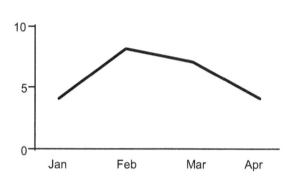

Line graphs are good for plotting trends.

Calculate rates

To make comparisons (for example, between departments or from year to year) it is often important to take into account the number of people working. For example, a large department may have more MSD cases than a small department, but also has more people employed. Calculating a *rate* will help tell you which department has a bigger problem.

To calculate a rate, take the number of cases, divide by the number of employees, then multiply by 100. The result is the MSD rate per 100 employees.

(A slightly more accurate, but more complicated method is to take the total number of new cases, divide by the *number of hours worked* that year, then multiply by 200,000.)

Comparing either cases or rates is correct, but each can give a different perspective.

Cases vs. rates

Here's an example to help show the difference between cases (or more generically, "numbers") and rates:

Last year, baseball player "x" had 20 hits, while baseball player "y" had 40 hits. Who had a better year?

40 vs. 20 may be an important difference, so is worth noting. But the number of times each player was at bat may tell a different story. Maybe "x" was up to bat 200 times and "y" only 60, resulting in batting averages of .200 and .333.

That's why you calculate the batting average — or to use the terminology from this kit — a rate. Rates help make comparisons.

More comments

- These analyses highlight the importance of keeping accurate records and being consistent for issues such as job titles.
- You can analyze the data for any relationship you want to compare, that is, generate dozens of tables and graphs to see what pops out.
- From a statistical point of view, the numbers of cases recorded on a log may be too small to know with certainty if differences are real or merely random. This type of analysis can still provide insight and give you accurate guidance, but don't get carried away by small differences.
- Likewise, be aware that the more miniscule the comparison, the more likelihood that differences will be random and not meaningful. For example, if you compare the number of wrist problems among Job B employees on the day shift versus the night shift, you might have such a small group that a minor difference either way can throw the apparent results way off. The comparison may still be worth exploring, but be leery of how much emphasis to place on results.
- This effect may vary considerably depending upon the size of the facility and the time period used. Larger plants and longer time periods normally have less variation.
- Also be careful of what you are looking at when making comparisons. Some data include wrist and arm problems only. Others include lower back problems and strains such as hernias. Or are you looking at a "rate" or a "number?" It can be confusing, so take your time.
- There are a number of commercially available computer programs to maintain these records, perform calculations, and print reports. Some are quite sophisticated and contain employee medical records, employee tracking systems, etc. These may be helpful for larger companies or facilities.
- Keep in mind that MSD cases and rates may go *up* at first, simply because an active ergonomics program *encourages* employees to report problems. Usually, the numbers eventually drop. But sometimes employers with best programs have recordable rates that remain high, simply because the higher level of concern and awareness can result in employees reporting more often than otherwise.

Hierarchy of cases

The total number of cases on the incident log should be the largest data base (assuming you have an incident log). This set includes everything. The injury and illness log provides a smaller data base, that is, just those that meet the government's criteria for recording. Cases with restricted days are smaller yet, followed by lost day cases. By definition, severity rises along this same hierarchy. All are worth studying. Each can provide a valuable insight and no single measure provides the "correct" view.

Expect Higher Recordable MSDs . . . at First

The experience at many companies is that the rates of recorded MSDs rise first when starting a systematic effort of prevention. This is good in many ways, since it means that you are identifying people who are having problems and getting them help at an earlier stage than otherwise (when the problems are much cheaper to treat and more likely to be successful). After a time, the rates usually drop. The overall pattern appears along the following lines:

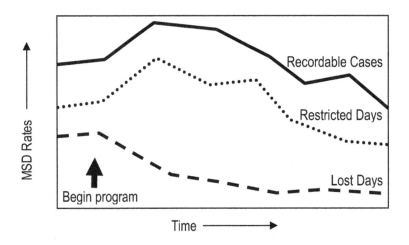

Typical trends

Surgery cases should be eliminated entirely. (If you have surgeries, you need to improve the system.)

Lost days typically fall almost immediately and remain low. Because these cases are the most serious and most expensive, this rate is often considered the most important one to track.

Restricted days usually rise then slowly drop off. It should be emphasized that because lost days drop and restricted days increase, this does not imply that an employer is "hiding" cases. On the contrary, this is the pattern you would expect from an active medical management program.

Recordable cases often rise because employees are doing what you are asking them to do — reporting problems early. This rate will also drop off as tasks are improved; however, in some ways it is OK for this rate to remain up for a long period of time (especially in high-risk industries like meatpacking or assembly), since it typically means that people are continuing to use the medical management system.

Subsequent peaks do not necessarily mean that conditions have worsened. The sample size of employees is usually small and from a statistical viewpoint there can be random variation.

Actions

Make sure senior management is aware of this possible effect, so that there are no surprises if it occurs. It may also be important to remove MSD cases from any bonuses that are based on injury rates.

Reward Systems

Bonuses based on recordable injuries

It is common for companies to base bonuses or other rewards on low recordable injuries. In the case of MSDs, however, this practice has a potential for conflicting with efforts to encourage employees to report problems.

As we have discussed, early recognition of MSDs is key, since the ailments can be treated at early stages with aspirin, ice packs, and restricted duty. Such conservative treatment is painless, inexpensive, and can result in complete recovery. If the employees wait too long, it is possible for surgery to be required, which may reduce the likelihood of complete recovery as well as being potentially painful to the employee and expensive for the employer.

Conceptually, it is good to reward workplace safety, since it sends a message that production isn't the only goal. However, the practice is often to use OSHA recordable injuries as the metric, which is not always best (see the preceding materials).

Resolution

One approach to resolving this dilemma is to remove the MSDs from the calculation of any bonus system. Thus the remaining injuries would be cuts, bruises, and falls which are related to safe behavior and hazard elimination.

Another, perhaps better, approach that has been adopted by a number of companies is to establish a type of management-by-objective (MBO) system whereby managers and supervisors receive rewards for the number and type of ergonomics issues resolved. These improvements can be administrative (such as improved task-specific training on best methods to do a job) or engineering (such as in-house modifications made).

Examples include:
- In a plant of a kitchen cabinet manufacturer, each supervisor was required to make one substantial ergonomics improvement each quarter (during the first two years of the effort).
- In other facilities, supervisors were rated based on the more itemized improvements the better.

I am aware of a major poultry processing company that established a numeric score for each task and then required the supervisors to make sufficient improvements to lower this score. This system took considerable effort to establish, but it had the advantage of being rather precise. The motivation was: No reduction in scores — no bonus.

Sample Ergo Log

Completed Items

Dept.	Task	Ergo Issue	Improvement	Status
12	Gluing table	Excessively high height of stand	Cut down	Completed 2/9
14	Manual Press	Repetitive arm motions to pull manual lever	Replace with air cylinder	Completed 6/3

In Progress

Dept.	Task	Ergo Issue	Best current idea	Status
42	Welding	Static load on shoulder from reaching out to weld the D frame	Provide arm rest	Ordered 9/5
14	Storage rack	High and low reaches to obtain parts from top and bottom shelves	Relocate heavy and high volume parts to middle shelves	Subcommittee formed, report due 10/10

Items on Hold

Dept.	Task	Ergo Issue	Idea	Status
12	Gluing	Static grip to hold glue gun	Add a strap to the gun	
12	Gluing	Reach for supplies	??	

Comments

1. In this word processing table, you can add additional columns that suit your needs, like "responsibility" to note the person who is to follow thru, or "cost" for the expense related to the improvement.

2. As the status of each item on the log changes (hopefully to "completed"), you can cut and paste the row in the table to a different spot to keep all similar-status items together.

3. For items that are in process, you can think of the column for "Improvement" as being "best current idea." The reason is that while items are still in the process of being evaluated and solved, the suggested improvement can change. This is good, since it generally means that people are thinking.

4. When a new idea is raised that seems to have merit, put it on the log, even if you cannot address it right away. Keeping it on the log helps prevent ideas from being lost or forgotten. You can also keep track of issues that are raised for which no ideas for improvement have yet been identified.

5. Status can be indicated with terms like: "active," "ordered," "work order submitted," "on-hold," etc.

6. If something doesn't work out, indicate it as "unsuccessful" and include a brief note as to why it didn't work.

7. This log can be set up in on the coordinator's computer and printed out when needed.

8. Use this log as a basis for reports, newsletter articles, presentations at conferences, and as a reminder to you that you are indeed achieving progress.

9. The Ergo log is an essential part of your written documentation

Describing an issue

When filling out an Ergo log, it can help to standardize the way you describe issues. This can enable other people to easily understand what you mean even tho you are using only a sort of short hand to save space on the form.

Altho there are many approaches to describing an issue, perhaps the best way is to briefly mention (1) the basic principle of ergonomics that is being violated, (2) the body part affected, and (3) the reason why. This three-phase description keeps you focused on the principle then helps lead in the direction of an improvement.

Principle affected		Body part affected		Why?
Awkward Posture		Hand		
Excessive Force	on/of	Arm	because	[describe in three
Repetitive Motions	the	Back	of	or four words]
Static Load		Neck		
Pressure Point				
Long Reach				
Awkward Height				

Examples:

Repetitive motions of the hand because of twisting caps onto bottles.

Repetitive, forceful motions of the arm to operate press.

Pressure point on thigh because of bar sticking out at workstation.

Helpful Terminology

Issue vs. problem

Altho it is common to speak in terms of *problems* and *solutions*, it can be better instead to use the terms *issues* and *improvements*.

The word *issue* is a more neutral term than *problem* and can help to keep people from being defensive (e.g., the manager of the area). Likewise, it can help avoid debates, especially early on in an evaluation when you don't really know if the situation is a "problem" or not.

More importantly, using the word *issue* helps facilitate the habit of making little improvements whenever possible. You don't want to wait until something meets everyone's definition of a "problem" before thinking about making changes. *Issue* helps keep the discussion away from an all-or-nothing mentality and promotes a good attitude of finding minor improvements that are appropriate for minor issues.

Improvement vs. solution

Similarly, the word *improvement* is often better than *solution*. *Solution* implies that the change is a perfect fix for all time, which is an incredibly difficult objective to achieve. Using *improvement* helps promote the concept of making a change that is feasible now, knowing that it can be made even better at some point. It is a never-ending process.

Ergonomic: a relative term

While we're at it, *ergonomically correct* is a horrible term. In reality, whether something is *ergonomic* or not largely depends on what you are comparing it to. A metal folding chair is more ergonomic than sitting on the ground. An old-style padded office chair is more ergonomic than a folding chair. A new office chair with multiple adjustments is more ergonomic than the old-style padded chair. (And if you've been standing all day, sitting on the bare ground can feel pretty good, too.)

Why These Distinctions Can Save You Money

All-or-nothing mentality

Many standard recommendations for "ergonomically correct solutions" can require unnecessary expense. There may be other equally valid improvements possible that are both less expensive and more effective. The all-or-nothing mentality of *problems* and *solutions* can lead to rash expenditures and a loss of common sense.

Phased-in changes

Often it is best to take a phased-in approach. An example would be to provide a small floor stand to enable a shorter person to reach high overhead. Then, the next time the work area is to be renovated and new equipment installed, redesign the layouts to eliminate the overhead reach altogether. Often, the costs of these types of changes are negligible when new equipment is installed.

Videotaping Jobs for Ergonomics Evaluations

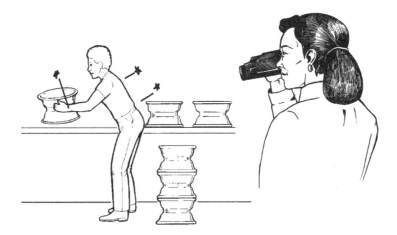

The video camera is the primary tool of the workplace ergonomist.

1. Before starting to tape be sure that all settings on the video camera are correct. It is usually a good idea to (a) check the battery power, and (b) turn on the clock or time code function.

2. It is a good idea to record the name of the job at the beginning of each separate job or task taped. This information can be written on a sheet of paper, then taped for several seconds.

3. If the people in the area are not familiar with you or why you are videotaping them, it is a good idea to introduce yourself and explain your objectives.

4. For most purposes, a few minutes of footage is all that is needed. Try to get at least a complete cycle of the task (for highly repetitive tasks several cycles). When in doubt, tape more. You can always fast forward the completed tape when you watch it.

5. Hold the camera still and take your time. It is sometimes helpful to place your right elbow firmly against your side and use your left hand under the camcorder to support it. In addition, place your eye firmly against the viewfinder cup. Do not walk with the camera unless absolutely necessary to record the task. If you must move or "pan" the camera, do so slowly to reduce recorded camera movement.

6. Begin taping with a whole body shot of the worker to provide perspective (be sure to include the chair or floor on which the employee is sitting or standing). Then move in to get more detail on the work being performed.

Ergonomics Consultant

Pros and cons

It is common to retain a professional ergonomist in the initial stages of establishing your program. The following provides an orientation to help you get the most from this relationship.

There are several advantages of involving a consultant for such assistance:

- The ergonomist has technical information that can help you in training your team and in solving problems.
- He or she may have seen many operations similar to yours and may know of good solutions.
- The consultant is an outsider, which more often than not results in instant credibility. Furthermore, the consultant is a neutral party, outside of organizational politics, which allows the ability to say things that might otherwise be awkward.
- The consultant may be skilled in facilitating brainstorming sessions and may have polished training programs.

The potential shortcomings of a consultant are that despite technical skills and academic training, he or she may not have experience in practical problem solving or no particular expertise in program development. You need to find someone who has actually fixed something on the workplace floor and not just taught anatomy classes.

Getting the most from a consultant

There are a number of steps you can take to help make sure you get the most from a consultant. The following is a hierarchy of consulting roles, which can help you and the consultant make sure you have the same expectations of the relationship.

Hierarchy of Consulting

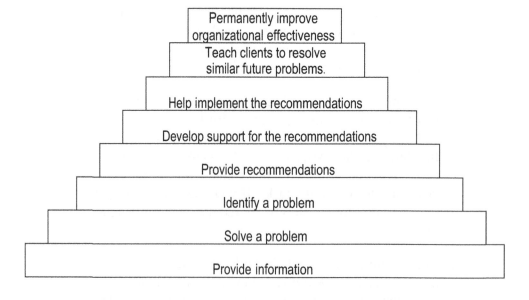

Many employers and consultants alike view the consulting task as the bottom rung or two in this hierarchy. But in my view, you should expect a lot more from a consultant than this.

If a consultant succeeds in getting people to *think* about site operations, the consultant usually has accomplished a great deal. As one plant manager once said to me, "Let me get this straight. You're a consultant who doesn't try to tell us what to do, but instead you try to get us to *think*?" I thought that was a superb way to regard the consultant's role.

In general, the best results usually happen when the ergonomics consultant does a combination of training and problem-solving. That way you get the best of both worlds: the consultant's ideas and comments, plus people at the site are better able to solve problems themselves in the future. And more importantly, the quality of ideas for improvements goes up when there is interaction between the consultant and site people, so it's not just the consultant making suggestions in the dark without knowing exactly how things are done.

Define the scope of work

It's helpful to define the specific actions or outcomes of the consulting project, for example:

- Provide a 3-day training session for TeamErgo with the objective of empowering the team to conduct task evaluations.
- Conduct four separate 2-hour sessions for supervisors on all shifts to explain the program and obtain their support.
- Meet with employees in the pilot department for 45 minutes.
- Conduct an analysis of OSHA 300 records and provide a short report of results in a PowerPoint format.

Sometimes you may need to meet with the consultant, provide a site tour, and discuss a variety of options before being able to define the exact scope of work.

Temporary help

Another common way to involve a consultant is as a temporary staff member. The ergonomist in this case serves all the functions as an in-house staff member. Sometimes the time commitment is full time for a matter of months and other times it is part time.

In either case, it is important to develop a process that involves permanent staff. You don't want your program to stop the minute the consultant leaves the door.

Developing a Written Ergonomics Program

Overview

The primary purposes of a written plan (whether for an ergonomics program or any other part of business) are to:

- clarify goals and plans, and
- communicate the program to others — for example, to senior management, new personnel, or regulatory officials.

In smaller operations, the written plan might only be a memo of one or two pages. In large operations with considerable problems, the plan may take a three-ring binder including a large section of appendices.

As with other types of business plans, the ergonomics plan should be reviewed and updated periodically. The program should reflect both the experiences gained in ergonomics and the changes that normally occur in the workplace.

Writing the plan

Writing a plan is not as difficult as it might appear. Often, it is simply a matter of putting down on paper (a) what is already occurring with the ergonomics program and (b) what those of you who are close to the program may take for granted about what you are doing. In some ways, the written plan thus serves to characterize the group's "collective consciousness."

You should also append copies of all program tools (such as questionnaires and forms) that you are using. Part of the idea is to keep in one place a copy of everything you use.

To many people, the hardest part of writing a plan is adopting the appropriate writing style. Thus, the following section is provided to help you adopt a business-plan style of writing.

Tips

Pretend you are writing a short report explaining your program to a newly hired plant manager — someone who currently knows nothing about what is going on, but who will be responsible for making sure the whole program works. You then write down a summary of things like:

- how you are structured
- what resources have been committed
- how employees are involved
- what training is conducted
- how you are focusing your efforts
- future plans and dates for engineering changes

Don't write down just what you think ought to be part of an ideal program. Describe only what you are actually doing or planning to do.

On the other hand, don't be reluctant to commit good, realistic plans to writing for fear of not being able to meet these plans. None of us always meets our own goals. Plans change as experience is gained. Nonetheless, it is still good to lay out a road map.

Sample written plans
On the following pages are examples of plans — written in two different styles — to help you see how you can best reduce your plan to writing. The first is a more informal approach written in a memo format. The second is a more formal business-plan approach. There are scores of ways to write a plan, but these should provide guidance and get you going in the right direction, especially for writing style.

Note that the two described are only a few of pages long each. For most organizations, a written plan does not need to be either long or elaborate.

Be aware that these sample plans are based on professional judgement and personal experience. They do not necessarily represent government interpretations of compliance.

Finally, remember that the following is only one way of organizing and describing a program. There are many other possible ways.

Memorandum

To: E. Whitney, President
From: C. McCormick, V.P. Operations
Subj: Ergonomics

The following is in response to your directive to have me develop
a plan to see what we need to do about ergonomics. All in all I
have determined that this could have some payoffs for us, if we go
about it right.

Organization

I formed a small team to help me. I've involved Frank (to get the
necessary maintenance support) and Sally (to serve as a
representative of the employees). I also spoke to both
supervisors about this plan, and that we will ultimately be
needing them.

Training

I'm gearing up on the learning curve pretty much as I did with the
Quality Process some years back, mostly by reading. I see that
the Chamber of Commerce has a speaker at the next meeting on
ergonomics, so I will attend.

I think ultimately I can pretty easily put together a series of
presentations that we can work into the employee safety meetings.
Most of this will be presented as learning "the rules of work" so
that we are better able to do our setups, but I'll make sure
everyone knows what to do if they are having any possible medical
problems.

I've found a good video that we can use as a start, but it appears
that most of this is down to earth, and I can turn what we will be
doing into short educational sessions (again just like what we did
on quality).

Communications

I'll add an agenda item to each of your staff meetings from now on
to keep everyone in management up to date. The safety meetings
will serve as the vehicle for everyone else.

Task Analysis and Priorities

There haven't been any back injuries or wrist problems (that
anyone's reported anyway).

In informal discussions, it seems the Doodad operation is still a
pain and what we should look at first. (I guess not much has
changed since you and I did that ourselves 10 years ago! And it's
really time we turned that from what we developed as a onesy-twosy
operation into something that supports our present volume.) And

we need to look at the times we weld overhead. From then on I think we may as well start at the beginning at the North Door and work our way down to shipping, and take a basic look at every step of our operation.

I found a pretty good worksheet that we can use to evaluate the jobs and use as the core of our efforts.

Job Improvements

I think right off the bat we should get some new anti-fatigue mats. The ones we have are a bit tired, to say the least.

Medical Management

I talked to the Doc over at the Feelwell Clinic to see what he had to say. As usual, they seem pretty up to date on this stuff.

He passed on to me that the nurse noted that when she treated Bill and Nguyen for the welding flash, both mentioned that they had some tenderness in their shoulders, which they said came from holding up the welding guns when they worked overhead.

Monitoring Progress

I think the best way to evaluate our progress is to just keep list of all the changes that we make. We should be able to make sure no one has any injury that will increase our workers' comp premiums.

SIX MONTHS TIMELINE

	Mar	Apr	May	Jun	Jul	Aug
Frank, Sally & I gear up	xxxx	xxxx				
Upgrade supervisors		xx				
Review ergo in safety meetings			xx	xxxx	xxxx	
Get more anti-fatigue mats	xx					
Review Doodad Operation			xxxx			
Develop plan for Doodad				xxxx		
Look at overhead welding					xxxx	
Start process-wide review						xxxx
Review and revamp						xx

Ergonomics Program — XYZ, Inc.

ORGANIZATION

Management Policy Statement

The XYZ Company recognizes the importance of ergonomics and the prevention of cumulative trauma as part of its overall safety effort. Promoting employee well-being and improving the production process from a human perspective is an essential part of doing business.

Accountability/Performance Review

The yearly performance review of all salaried production supervisors will be modified to include accountability for ergonomics. The modified system will be in place by November of this year in time for the January review.

Coordinator

A position has been created for a full-time ergonomics coordinator. The job description is attached in Appendix A.[*]

Committee

TeamErgo consists of the following representatives:

- plant superindentant (chair)
- director of engineering
- first-shift nurse
- ergonomics coordinator
- union president
- union steward for maintenance department

Additional ad hoc members of the committee are appointed as needed to assist in special projects. The purchasing manager also attends as needed.

The committee meets on the first Tuesday of every month at 3:00 PM. Minutes are taken and action items noted.

Department Teams

Temporary ad hoc department level teams are created to assist in problem solving when jobs are targeted in that department. These teams are modeled after the Quality Improvement Teams that address similar issues. Teams consist of:

- two or three volunteer employees
- department engineer
- TeamErgo member

---

[*] *Actual appendices are not included in this sample plan.*

Financial Resources

A budget of $35,000 has been committed for ergonomics projects. This budget serves to fund small projects for which return on investment cannot clearly be calculated. Additional funds for projects with ergonomic benefit are available as part of normal capital expenditures.

Maintenance Personnel

Two full-time maintenance personnel are dedicated to ergonomics and process research and development. These personnel report directly to the director of engineering.

Consultants

The company has retained ongoing services of two consultants to assist in program development:

- Acme Ergonomics, Inc., professional ergonomists
- Dr. K. Feelwell, M.D., occupational physician

Employee Involvement

- Employee representatives are members of TeamErgo (see above).
- Volunteer employees are members of the Departmental Teams (see above).
- Interviews are held with individuals whenever specific jobs are reviewed.
- Ergonomic issues are raised in department Quality Improvement Team meetings.

TRAINING

TeamErgo

At the initiation of the program, a four-day training session was held for the committee. All members of upper management attended the first half-day of that session. Handouts are attached in Appendix B.

Supervisors

All supervisors attended an initial three-hour session. One-hour refresher sessions are held annually as part of the "Management Skills Training" program. TeamErgo members provide instruction. Handouts are attached in Appendix B.

Employees

All employees attended an initial two-hour session on ergonomics. One-hour refresher sessions are held annually. TeamErgo members provide instruction. Handouts are attached in Appendix B.

The "Skills Building" training program has been modified to include basic principles of ergonomics as part of new-hire orientation.

Maintenance Personnel

All hourly and salaried maintenance personnel attended an additional three-hour session. Additional sessions will be held on an as-needed basis. Handouts are attached in Appendix B.

COMMUNICATION

Employee Updates

Employees are kept abreast of developments in the following ways:

- Employees in areas targeted for ergonomic intervention are notified as initial job analyses are conducted. Feedback on status of planned improvements in affected areas is provided during the appropriate departmental Quality Improvement Process meetings.

- Articles on overall progress in the ergonomics program are included in the quarterly employee newsletter.

- Before-and-after photos and testimonials by employees concerning job improvements are included in the Quality Improvement bulletin board.

Plant Communication

Overall coordination of ergonomics activities is accomplished in two ways:

- TeamErgo meetings
- Plant Manager staff meetings (Director of Engineering's report)

WORKSITE ANALYSIS

Priority Setting

Background assessment to gather information needed to set priorities is conducted in five ways. TeamErgo reviews results from each approach and decisions are made to select target jobs for problem solving for the upcoming quarter. The Planning Guide used in this process is attached in Appendix C.

OSHA 200 Analysis

Injury/illness statistics are calculated and graphed quarterly. As part of this process, CTD cases are evaluated separately.

Workers' Compensation Cost Analysis

Workers' Comp costs are broken down by department, and where possible, by job on an annual basis. Costs are itemized separately for both acute and cumulative trauma.

Walkthru Surveys

Background walkthru surveys are conducted to identify tasks that may need further, more focused review. Brief discussions are held with employees at their workstations as part of this analysis. These surveys are accomplished in two ways:

- Each month two members of TeamErgo survey a different department. A primary goal is to review tasks that may not have been in operation on previous surveys. The checklist used for this purpose is attached in Appendix D.

- The Ergonomics Consultant also conducts a brief walkthru of selected departments to help ensure that all issues have been identified.

Renovation and Expansion

Whenever a process or work area is scheduled for change or expansion, a walkthru survey is conducted and the Quality Improvement Team consulted to determine if opportunities for ergonomic changes are present.

Problem-Solving

Targeted jobs are addressed in a two-tier approach:

- Problem-Solving Worksheet

  The targeted task is analyzed using the worksheet attached in Appendix E. After initial facts are gathered, possible improvements are brainstormed in a meeting involving the Department Team and members of the TeamErgo.

- Engineering Study

  Tasks that cannot be adequately addressed in a simple brainstorming meeting are referred to the engineering department. The Department Team is still involved in the process. The Ergonomics Consultant is also involved if appropriate.

Employee Discomfort Survey

Employees in targeted areas are administered an anonymous Physical Assessment survey. A copy of the survey form is attached in Appendix F. This survey is only used when improvements are sought for a specified work area.

Medical Management Job Analysis

Jobs are also reviewed for the purposes of (a) medical management and (b) meeting obligations of the Americans with Disabilities Act. These analyses are described below under Medical Management.

JOB IMPROVEMENTS

Implementation

Improvements are executed in four ways:

- Work Orders — The Ergonomics Coordinator or Department Supervisor may write a work order for simple changes as part of the problem-solving process described above. Ergonomic issues are identified by checking as appropriate either the "quality" or "safety" box on the form for higher priority.

- Capital Expenditures — TeamErgo or Engineering Department may submit a request for capital expenditure using standard procedures.

- Human Resource Changes — TeamErgo may work with the Human Resource Department to institute changes that involve staffing, task reallocation, work/rest schedules, and similar issues.

- Purchasing — The Purchasing Department reviews appropriate purchase decisions with either the Ergonomics Coordinator or TeamErgo. For major purchases a Quality Improvement Team from the affected area is also contacted and a review of ergonomic issues is performed as part of that activity.

TeamErgo keeps an action list of on-going projects, which includes an implementation schedule for all items. The current schedule is attached to this plan in Appendix G.

MEDICAL MANAGEMENT

General Policy

The XYZ Company recognizes the importance of a comprehensive medical management system as part of the overall program to prevent CTDs.

Medical Providers

An occupational health nurse is staffed on both production shifts. Further medical support is provided by the Feelwell Occupational Medicine Clinic.

Periodic Walkthru

One of the nurses is a permanent member of TeamErgo and participates in walkthru surveys as part of that effort. The other nurse conducts a walkthru of one department per month in conjunction with the department supervisor.

Restricted Duty - Job Matching System

A formalized system is used for evaluating suitability of jobs for employees with medical restrictions. Medical providers fill out the Employee Physical Restrictions form (see Appendix G) and TeamErgo members provide information on potentially suitable jobs using the Physical Job Requirements form (see Appendix H). The appropriateness of the job for an individual's personal restrictions is determined by matching the two forms.

Active Medical Surveillance

On a pilot basis, the nurse are providing upper extremity and lower back physicals biannually to the employees in Department 12, the highest injury department. Based on experience with these physicals, a determination will be made by June of next year about whether to expand the use of these physicals in other departments.

Protocols for Evaluation and Treatment

The protocol for evaluation and treatment of MSDs is based on the OSHA Meatpacking Guidelines.

MONITORING PROGRESS

Management Review

The overall program is formally reviewed quarterly at the Plant Manager's staff meeting. Modifications in the program are reflected in this written plan.

Injury/Illness Trends

Summary reports are written quarterly and annually.

Ergonomics Log

The Ergonomics Coordinator maintains a log of all completed projects. The completed log is attached to this plan as Appendix H.

Special Study

Line A in Department 12 was the target area for a major ergonomic improvement. Employees on this line are being evaluated as part of the nurses' Active Medical Surveillance and results before and after the changes are being compared. Additionally, the Ergonomics Consultant has measured risk factors before and after the changes.

Program Development Worksheets

Several worksheets are included in this kit to aid you in developing and writing your plan.

Listing Outcomes

Often it can be helpful to list the objectives or outcomes of certain stages of your program. This approach has the advantage of focusing on results rather than the steps of getting there.

It may seem self-evident and simplistic to write lists like these, but the act of writing makes your goals communicable to other people and serves as a basis for discussion. When you don't write down objectives, it is easy for people to make wrong assumptions that lead to confusion and misunderstandings.

Some examples are found below:

By the end of the startup period we will have accomplished the following:

Organization

- The plant will have an organizational system in place to address ergonomics issues, with clear responsibilities and accountability.

- All managers and supervisors will understand the basic ergonomics issues in their work areas, the general strategies for making improvements regarding those issues, and the importance of implementing these improvements.

- An Ergonomics Committee will have been formed and will be prepared to coordinate ergonomics activities after the ergonomist departs.

Training

- The ergonomics subcommittee will have received the appropriate training and guidance to be capable of performing basic ergonomics task evaluations and will have performed several of these evaluations.

Task Evaluations

- A process will have been created for ongoing solicitation of employee concerns and ideas.

Task Improvements

- Safety Committee members and other plant personnel will have implemented a number of improvements and have plans for additional improvements.

Chapter 8
Creating Change

You can create change in your organization, even if you have no formal authority or command a budget. You can persuade and be persistent, and you can succeed.

Learning how to create change is important, because a huge part of the workplace ergonomics process usually involves selling ideas and getting people to accept new ways. Unfortunately, this topic is almost never taught in classes on ergonomics and some people are unsure of what to do. The following material provides some brief pointers for you.

Being a change agent

Change agents are always self-appointed. You cannot assign someone to this, so a good start is to think of yourself as one.

The best change agents are diplomatic in working with other people, willing to ask tough questions, and thick-skinned enough to face criticism and resistance to change. Good communication and social skills are obviously important talents to have.

Don't worry if you feel that these skills might not be your strongest aptitudes. They are not exactly prerequisites for the job; rather, they can be helpful to you. You probably are capable enough already, and with some reflection, you can improve the skills you already have. The point here is for you to think about the change process a bit, which can help you become better at it.

How to influence people

Change is a component of all the modern techniques of the advanced workplace: improving customer service, the quality process, lean manufacturing, re-engineering, etc. Moreover, all of these systems reinforce one another in promoting and managing change.

There are countless books, training seminars, and information about creating change, but one of the best is still *How to Win Friends and Influence People* by Dale Carnegie. This is a classic self-help book that everyone should read, whether you are a change agent or not. Ironically, the lessons have more to do with you being polite and respectful than with learning any devious or manipulative techniques.

Carnegie provides a set of rules along with stories to illustrate their use. Examples of his homey advice are:

- To get the best of a situation, avoid arguments.
- Always listen to others' opinions and never tell anyone they are wrong.
- Admit it if you are wrong.
- See the other person's point of view.

His simple prescriptions offer a good, down-to-earth way for you to think about how to go about creating change. If you are able to incorporate them into your day-to-day behavior, you should end up being very effective in accomplishing your goals.

Understand the change process

Change must be addressed at two levels: (1) each individual and (2) the organization as a whole. This distinction can create the age-old dilemma of trying to decide whether to create change from the top down vs. the bottom up.

In reality, you cannot affect one without the other. If individuals begin to change, but the organization does not, the individual changes are easily frustrated and stifled. On the other hand, if the organization starts to change from the top down, but the individuals do not, then nothing happens.

Consequently, it cannot be an "either/or" question. You must work on both simultaneously.

Resistance to change

It is normal for there to be resistance to change and you should anticipate at least some challenge. Change can be stressful. People can fear the unknown, or fear that the new way may not be better.

Part of the problem is that we have inherited a legacy of not involving employees in planning. Far too often in the past, employees were just told what to do, without any effort to explain or convince them of the need. Or worse, employees showed up on a Monday morning to find dramatic changes in their work areas. Not only were the employees not involved, no one even had the common courtesy to inform them that the change was planned. There may be a climate of mistrust, which makes it even harder.

No wonder there is resistance to change! But don't be overwhelmed. There are plenty of examples of how people just like you have been successful. Everything in this book is geared to help you implement change.

Furthermore, ergonomics is often an easier sell than many other topics, because the issues are so immediate and tangible to many people. An improvement like fewer long reaches appeals to most people and provides a direct reward for their support. It's not like improving the quality of products and services, for example, where the benefits are important, but indirect and often not very immediate to many employees.

Individual Change

Many ergonomics improvements are so much better than past practice that getting people to accept a change or use a new technique is not a problem. But if the new way violates old habits, there can be some resistance. And on occasion, you can run into militant resisters, who stand with their hands on hips, look you sternly in the eye, and ask, "Why do I have to change?"

The basics

Train, involve, and empower — You start with education by explaining what, why, and how. For instance, people who work with adjustable equipment need to be told how it is adjusted and the goal of using the equipment. Participation in the process of identifying issues and possible solutions creates ownership of ideas and helps people to buy into the process, rather than having it dictated from above.

Provide practice — Remember that instructions are not enough. People must practice a technique for it to become a habit.

Acceptance of new methods or tools

It is common in the ergonomics process to be in the position of trying to get people to adopt a new method or piece of equipment. Changing individual habits is not an easy mission and can take a long time. While some individuals will adopt everything you say and are appreciative, others may never do what you think is best.

There are many aspects to this part of the change process, but the following are a few tips to help you:

Muscle memory — One of the reasons why it can be difficult to get acceptance is the phenomenon of "muscle memory." This term refers to a physical skill that we have mastered so thoroughly that we no longer have to think to perform it. In the workplace, it is common for individuals to have done certain tasks for so long that any change can feel awkward, even if it is clearly better.

It helps to explain this factor to employees, so that they can anticipate that it may take them a while to give the new method a fair test. You can't usually try something for a few minutes and decide whether it is an improvement or not. Once again, this is part of communications and grooming people's expectations.

Try it, you'll like it — Sometimes, you have to cajole people into trying a new method or piece of equipment for a while. The length of trial must be long enough to overcome old habits.

Change everything — Sometimes it is easier to disrupt everything, so that everything is new and the task has to be learned all over again from scratch. Bear in mind that production standards may need to be relaxed for a time to allow people to get up to speed.

Change for the sake of change — Some companies have learned that it is good to make changes at regular intervals just to keep people from getting into ruts. If people don't understand the objectives for making these changes, it can seem a bit bizarre to keep changing layouts and moving work areas around, but with proper communications there is merit to this practice.

Condition of employment — Eventually, if you are certain that the new method is safer and you still have resistance, you may need to make the method a condition of employment and apply your disciplinary system. You need to be sure that the issue you are dealing with is not one of personal preference or that there is some legitimate reason why an employee is resisting the change. But there is no reason why these issues should not be treated like any other safety rule.

New hires — Make sure new hires are started off on the right foot. They generally don't have old bad habits to overcome.

Behavior-based safety — The growth of behavior-based safety programs in recent years has provided a rather complete strategy for dealing with individual change. Some organizations may find it valuable to incorporate these techniques in their efforts.

It all adds up

Critical Mass — Many experts on change refer to this one-on-one convincing as being "missionary" work. You convert individuals one at a time to think in terms of good ergonomics. Eventually, you will convert enough people to create a critical mass that accepts the new approach. It becomes a tipping point so that the organization as a whole changes.

A few more tips

Sincerity helps in quite a number of ways. You can't fake this. People can tell. But being sincere can help you to achieve your goals.

Photograph smiles — Often you are in the workplace with a camera, taking photos of workstations and tasks. You are then in a good position to take photos of people just being themselves. With a bit of camaraderie you can get them to smile naturally.

If a photo comes out well, use a color printer and print out one copy to give to the individual and another to frame and hang in gallery of smiles in some appropriate spot, like the training room or cafeteria. The point is that you can make friends this way and build support.

Organizational Behavior

Gaining credibility

A first step is to change something and start building some credibility. It is difficult to convince people with sweeping promises about all the nice things that you *want* to do. It is much more effective when you can point to things that you have already *accomplished*. This can work even if the changes you have made are modest, as long as you describe the changes as modest and don't overstate them.

Communications

How you communicate is essential.

Vision — You need to provide people with an image of what the change will be and how they fit in.

Rationale — You need to explain "why."

Plans — You must provide information on when, where, and how the change will be implemented.

None of this needs to be complex or grandiose. You can make simple statements at meetings like "We are going to make a concerted effort to find smarter ways of doing heavy tasks, starting in Dept. XYZ this Thursday."

Buzz words

Most organizations have certain buzz words, that is, catch phrases that capture the immediate needs or goals of the organization. Use them, as appropriate. You need to think about what else is going on in your organization and shape your efforts to fit those. You want to be able to take advantage of momentum that has been created in those efforts and, in turn, contribute to that momentum with your own energy.

Tap into self-interest

Think of your audience's self-interest and explain your goals in terms of their self-interest, that is, what's in it for them. This is not being phony or manipulative (assuming you're not leading people on). It is good communications.

To employees, you want to emphasize how ergonomics can result in less wear and tear on their bodies. To managers, you normally would stress how ergonomics can result in lower costs and greater efficiency. To supervisors, you might say something about reducing their headaches, such as a problem operation that as been creating both downtime and injuries. To engineers there might be something about identifying a challenging problem that might raise their interest.

The point is that if you want to lead people somewhere, you need to show them that what *they* want coincides with what *you* want. If you can show them a specific step to achieve their objectives, they may well take that step.

Use their experience as reference point

It can be harder to convince people of something if they view it as too foreign or remote from their lives. Conversely, if you make things seem familiar, people can accept it more easily. Like every other aspect of good training, you want to make your message seem accessible.

In part, this is why it can help to explain that ergonomics is really nothing new, as mentioned in Chapter 6 *The Workplace Ergonomics Process*. Or as another example, it can help to use phrases or examples from your audience's own traditions.[*]

Celebrate milestones

One common problem is that people within an organization all too often cannot see the extent of the changes that are occurring around them. They are sometimes so close to the trees that they cannot see the forest. If this is the case in your facility, you must find ways to take a step back and list accomplishments. Celebrating milestones is one way to do this.

Look for opportunities to plan some event to mark achievements. For example, provide lunch when you meet a goal in reducing lost days, or in implementing a certain number of improvements. Such events can cause people to get away from day-to-day frustrations, reflect on progress, and thus gain a sense of momentum.

Dust off your lists

It is common during training sessions for supervisors and employees to hear something like the following from one of the participants, "I've been submitting suggestions for years on items like this and they've all been turned down. Why should the company listen now?"

A good answer is, "Dust off your lists, add the word ergonomics, and resubmit them. We have a team now and an active process for this purpose. Your ideas may fall on more fertile ground now."

Final thoughts

- Try to engage resistors not stifle them. Consider putting some of the active ones on your team. You might be able to turn that energy in your favor.
- Remember to give credit to originators of ideas and teams that have worked on projects. That helps build the forces on your side.

---

[*]In New England, I usually promote ergonomics as "Yankee Ingenuity," since it taps into the self-image of people there. Down in Dixie, I often refer to "Southern Engineering" or "Shade Tree Engineering." In China, I have invoked the legendary figure of Zhuge Liang, a strategist of the Three Kingdoms period, known for his clever thinking.

Ways to Get Management Commitment

In some companies, senior management leads the way in promoting good ergonomics. In other companies, it may be necessary for you to explain and convince top management of the importance of your efforts.

The following list contains ideas on how you might go about getting management commitment for your ergonomics plan. Organizational cultures vary, so you need to adapt these ideas to your situation.

Presentations

Usually, one of the initial steps you take is making a formal presentation to your management group. You need to review your plan and explain why ergonomics is important. By laying the groundwork now, you pave the way for future support. Often, very little else can be accomplished until you have taken this first step.

It helps to show the costs of doing nothing and the possible benefits of a good program. Make sure the upper management group knows the direct and indirect costs of workers' compensation for your organization, and that many companies have been able to reduce these costs. Make the point that companies *can* control workers' compensation costs and that they are not an unavoidable cost of doing business.

Provide an overview of ergonomics

Use plain language and try to show that the field makes common sense.

Show videos of problem jobs, or give the management group a tour of problem work areas, pointing out specific issues. Sometimes, upper management simply isn't aware of problems at specific workstations (even in their own offices, at their own desks).

Address how ergonomics can make business sense in a way that fits your specific business. Talk about the efficiency aspects of ergonomics, since typically, if managers have heard about ergonomics, it has only been in reference to government regulations. Use materials from this kit to create ideas and customize to your situation.

Planning

Structure your plan to fit the business vision. List your company's strategic goals and initiatives, then show exactly how your ergonomics plan fits into these strategies and furthers these goals. Then make a handout and/or presentation slide that shows how your plan can help your management achieve its vision.

Note that some goals are explicitly listed in mission statements and policies. Others are more implicit or are simply common knowledge in your organization.

Also, some goals might be lofty, such as "Support and Inspire Each Other," while others are nuts and bolts, such as "reduce costs 10 percent in the next fiscal year." Despite these variations, it is likely that there are ways in which your ergonomics program supports these efforts (and if not, maybe it should). The point is to explain these parallels to the powers-that-be.

Write your plan

A written plan helps prove to your management team that you have thought the issues thru. They may not read your plan line by line, but typically they like to see that you have put effort into it.

If necessary, find a champion, that is, someone in top management who you can talk to separately and who will argue your case in the press of day-to-day business life.

Provide a list of recommendations

Some senior managers can be very much in favor of making changes of the sort we've been covering. However, they may not be familiar enough with operations at the level of detail that you may be, so they need to see specific ways to make changes. Your team can do a number of work area evaluations, the compile a list of the most meaningful ideas.

Show results

No one likes to waste money on things that are not effective, so you should show that the investments in ergonomics are well spent. Cost reductions are typically the most powerful evidence for business people, but even if cost-saving data are not available, there are other ways to show results. Most good managers recognize that not everything can be documented in dollars and cents.

For example, provide before-and-after evaluations, or describe the problems fixed at a particular workstation, or the number of workstations evaluated and improved. Even if you cannot yet translate these improvements into cost savings, at least you can show that the ergonomics process has achieved positive results in its own terms.

You can also provide results of employee surveys that show reduced discomfort or increased morale. Another approach is to provide plain language testimonials from people; for example, a note from Mary in Dept. 32 stating that she no longer suffers neck pain because of a low work surface.

Compare

Benchmark other companies. Managers are almost always interested in trends among other companies, in particular, among leading companies who have already established good ergonomics programs. You can make the case that the big showcase companies are doing ergonomics now, so you should, too.

Threats work, too

Highlight the multimillion dollar OSHA fines, negative publicity, or the litigation that some companies have had to undergo. Don't overdo this, but if you think you are vulnerable, you need to say so.

Atmosphere of Innovation

Experience

Many of the best cases of success in ergonomics have come when the organizational climate is one that supports innovation. In some cases (such as Case Example 3 in Chapter 4), ergonomics fell on very fertile ground.

In other cases (such as Case Examples 1 and 2), it was the ergonomics effort that paved the way for organizational change. In the process of butting heads to make ergonomics improvements, the team broke down many of the old barriers and created an atmosphere where innovation became normal and expected. In these examples, the ergonomics effort led to other changes, such as in product quality and customer service.

In either situation, it is so helpful to have an organizational culture that supports changes that it should become one of your broader goals. An atmosphere of innovation can help you, and conversely, you and the ergonomics process can help develop it.

New rules

It helps to adopt a set of "rules" that set the tone for change. A sample list is provided on the next page. The idea is to make it known that there are now new expectations and that people are supposed to make suggestions.

You can see by reviewing this list that often people need to be provided with some guidance. For example, some people might expect that every idea they suggest will be implemented without question. This is an unrealistic expectation and you need to let them know that there is more to it than that. The entire list helps to layout a whole strategy: Not every idea works, but that's OK because the point is to raise lots of ideas.

Poking fun

An effective tool in breaking down opposition is humor. Page 223 provides a list of statements that people tend to make when they don't want to try something. By listing these quotes, when people find themselves saying them, they often stop and laugh at themselves, then give up.

Upside-down pyramid

Page 224 provides a visual concept for organizational change. The traditional view of the organizational chart is that the boss is on the top, who tells people below what to do, the supervisors tell the employees, etc. Of great significance, there is no place for customers.

In the upside-down organizational chart, the customers are on top (which, after all, is the whole point of the enterprise). The boss is at the bottom of the chart with the job of supporting the managers, making sure they have the capability and authority to do their jobs. The managers do the same for the supervisors, who in turn do the same for the employees.

Management's job is to push decision-making "upward" and to make sure that decision-making is competent.

NEW RULES
For Innovation

1. Everyone is expected to come up with ideas for improvement.

2. Raise as many ideas as you can. Whoever thinks of the most ideas wins.

3. Teams identify the best ideas. Two heads are better than one, and 12 heads are better than two.

4. Thou shalt brainstorm. Say whatever crosses your mind. Play off other peoples' ideas. Harebrained ideas are required. Avoid focusing on reasons why an idea won't work, rather keep generating ideas.

5. Watch videotapes of operations. This causes you to focus and you will see things in new light. Do so in a meeting room to permit everyone to be involved and without interruption caused by plant noise.

6. Do not feel discouraged if an idea doesn't make it past the discussion stage. Fewer than one of ten ideas that are raised are actually feasible.

7. Do not expect ideas to work right immediately. Instead, expect that it will take time and usually a bit of trial and error to get something to work right.

8. Understand the difference between (a) a snag, which can be overcome, and (b) a fatal flaw, which totally kills the concept. Do not give up at the first snag.

9. Experiment. If you don't know if something will work or not, test it. Rig up a trial. Build a prototype.

10. It is OK to expect *some* failures. Thomas Edison failed thousands of times at making a light bulb work until he found the right combination of materials.

Thirteen Ways to Kill a Good Idea

1. It will cost too much.

2. We don't have the time.

3. We've tried that before.

4. We've never done that before.

5. It won't work here.

6. I've never heard of anyone else doing it that way.

7. We did all right without it.

8. This is the way we've always done it.

9. Management won't go for it.

10. Employees won't go for it.

11. It's not our responsibility.

12. Why change? It's working OK.

13. It might work — and then we'll have to change.[*]

This page available electronically.

---

[*] Adapted from an unknown source.

The Upside-Down Organizational Pyramid

Traditional Organization Chart

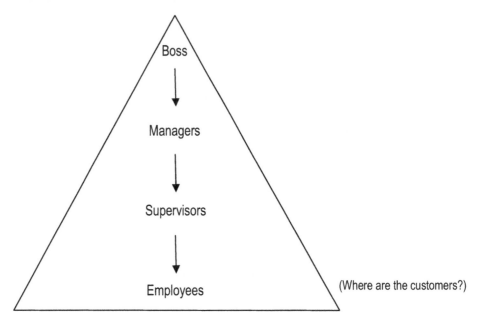

Boss

Managers

Supervisors

Employees

(Where are the customers?)

Customer-Oriented, Employee-Oriented Organization Chart

(The Upside-Down Pyramid)

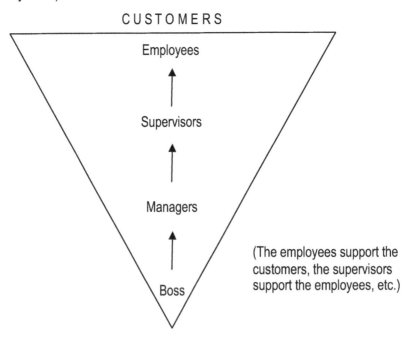

CUSTOMERS

Employees

Supervisors

Managers

Boss

(The employees support the customers, the supervisors support the employees, etc.)

This page available electronically

Chapter 9
Government Standards and Guidelines

There are a number of standards and guidelines that have been developed in recent years that can provide guidance for your program (if not requirements):

United States

- Federal OSHA's *Ergonomics Program Final Rule* (rescinded by Congress)
- OSHA 5(a)(1) enforcement practice
- Federal OSHA's *Ergonomics Program Management Guidelines for Meatpacking Plants*
- Other Federal OSHA industry-specific guidelines
- California OSHA's *Repetitive Motion Injuries* standard.

Other Countries

The U.S. is certainly not alone in attempting to develop guidelines on the workplace ergonomics process. Canadian provinces have developed a variety of requirements, the most comprehensive of which are those of British Columbia. Australia has a set of broad requirements and guidelines. Above all, Sweden has developed far-reaching directives and rules.

It can be helpful for Americans observe some foreign approaches. It is not the intent here to survey all countries, rather simply to provide a sense of events outside the U.S. Simultaneously, the materials in this book may be of value to people thruout the world.

Thou Shalt Have a Process

With few exceptions, these documents do not address specific issues like heights of workstations or shapes of hand tools. Rather, they encourage management to set up programs to systematically address ergonomics. In other words, these rules state "thou shalt have a process."

RMIs, MSIs, CTDs, OOS, MSDs, and VDTs

Each government has used a different term for the ailments that are called musculoskeletal disorders (MSDs) in this book. As described in Chapter 3 *Understanding MSDs*, these terms are equivalent: repetitive motion injuries (RMIs), musculoskeletal injuries (MSIs), cumulative trauma disorders (CTDs), and occupational overuse syndrome (OOS). Also, note that some countries have standards related to computer ergonomics (omitted here). These rules often refer to "video display terminal" (VDT), which is an obscure term that means "computer screen."

United States

Outline of OSHA Ergonomics Program Standard

2000

The U.S. Occupational Safety and Health Administration (OSHA) promulgated a standard in 2000 that required employers to establish programs to prevent MSDs. However, shortly thereafter, this regulation was rescinded by an act of Congress. The fact that Congress felt it was necessary to take such an unprecedented action to block a regulation shows the depth of controversy regarding this issue.

Despite the fact that the standard is not in effect, it is helpful to review some key components. The core of the standard was a set of provisions that are outlined below:

(*Excerpts*)

Management leadership
- effective MSD reporting system
- prompt responses to reports
- clear program responsibilities
- regular communication with employees about the program

Employee participation
- early reporting of MSDs
- active involvement by employees
- implementation
- evaluation
- future development of program

Job hazard analysis and control
- a process that identifies, analyzes, and uses engineering, work practice, and administrative controls to control MSD risk factors to the extent feasible
- evaluates controls to assure that they are effective

Training of managers, supervisors, and employees
- your ergonomics program and their role in it
- the recognition of MSD signs and symptoms
- the importance of early reporting
- the identification of MSD risk factors in jobs in your workplace
- the methods you are taking to control them

Program evaluation
- regular reviews of the elements of the program and of the effectiveness of the program as a whole
- reductions in the number and severity of MSDs
- increases in the number of jobs in which MSD risk factors have been controlled
- reductions in the number of jobs posing risk of MSDs to employees
- correction of any deficiencies in the program

Other Requirements and Guidelines

General duty clause

OSHA does not need a standard to cite a company on ergonomics issues. In the legislation that established OSHA in 1970, one of the paragraphs stated that every employer has a duty to provide a workplace free of recognized safety hazards. This paragraph is referred to as the "general duty clause" or by its paragraph number in the law — 5(a)(1). OSHA has used this clause in a wide variety of cases to cite employers for problems where no specific standards have been adopted, including a number that involved MSDs.

OSHA's burden of proof — Administrative court rulings have confirmed that OSHA may cite employers for MSDs using this general duty clause, but that OSHA must show certain facts. OSHA's burden of proof is along the following lines:

- MSDs and risk factors exist in the workplace.
- Feasible improvements are available and that they have likely utility in reducing injury. Note that "feasible" means both technological *and economic* feasibility; OSHA must be realistic.

The case law is a bit complex from a lay perspective. If you are affected, consult an attorney who specializes in OSHA matters.

Settlement agreements — Some of the largest fines that OSHA has ever levied on employers have been related to MSDs and the agency continues to issue citations. Settlements usually involve written agreements whereby the employer commits to taking certain actions.

Some of these actions have to do with fixing specific problems in that workplace, including making certain changes in equipment and work practices. But key elements of these settlement agreements usually involve a commitment to establish a formal ergonomics program.

Finally, it should be noted that OSHA's practice has been to recognize the existence of a good faith ergonomics program. "We want to see employers going in the right direction," is a phrase that some OSHA personnel have used. (Not all OSHA inspectors and administrators agree with this philosophy. Some view their jobs as "poking holes" in company programs, no matter how good those programs happen to be. But in general, having a good workplace process for dealing with MSDs can mitigate problems OSHA finds on site.)

OSHA meatpacking guidelines

One of the landmark actions in the history of workplace ergonomics was the development of OSHA's *Ergonomics Program Management Guidelines for Meatpacking Plants* adopted in 1990. The meat industry had been suffering from extremely high rates of MSDs, which prompted a series of high visibility OSHA citations and fines, then subsequently these guidelines. The guidelines are voluntary and were developed in cooperation with the American Meat Institute (AMI), the trade association for the meatpacking industry.[*]

In outline form, the guidelines encourage Management Commitment, as evidenced by:

- a written program
- employee involvement
- regular program review and evaluation

The guidelines are also list four program elements:

I. Worksite Analysis
II. Hazard Prevention and Control
III. Medical Management
IV. Training

These guidelines were instrumental because they were the first time that many aspects of the workplace ergonomics process were published. For example, for years, the overview of medical management of MSDs contained in this document was the only widely available guidance for occupational health nurses. Furthermore, the meatpacking guidelines set a pattern for development of similar actions in other industries.

Guidelines for nursing homes

In 2003, OSHA developed a set of voluntary guidelines for nursing homes, an industry that has been affected by high rates of lower back injuries from lifting patients. These guidelines differ from the meatpacking guidelines in that there is more detailed information on solutions to the specific task of lifting patients, plus case examples of how certain nursing homes saved money thru the use of ergonomics.

---

[*] I represented the AMI in working with OSHA to develop these guidelines. The trade association had, in fact, previously developed its own internal guidelines, which I drafted, that were equivalent to the content of OSHA's guidelines, altho markedly different in writing style.

Guidelines for retail grocery stores

In 2004, OSHA developed a similar set of guidelines for grocery stores. An outline of the program elements is:

Provide management support

- Develop clear goals
- Express the company's commitment to achieving them
- Assign responsibilities (training, job analysis, etc.) to designated staff members to achieve those goals
- Ensure that assigned responsibilities are fulfilled
- Provide appropriate resources.

Involve employees

- Submit suggestions and concerns
- Identify and report tasks that are difficult to perform
- Discuss work methods
- Provide input in the design of workstations, equipment, procedures and training
- Help evaluate equipment
- Respond to surveys and questionnaires
- Report injuries as soon as they occur
- Participate fully in MSD case investigations
- Participate in ergonomics task groups

The guidelines continue using elements entitled: Identify Problems, Implement Solutions, Address Reports of Injuries, Provide Training, and Evaluate Progress.

Other industry-specific guidelines

These brief outlines and descriptions of industry-specific should serve to provide some understanding of OSHA's expectations for key elements of ergonomics programs. There are additional industry-specific guidelines, some initiated by OSHA and others initiated by the industries themselves in cooperation with OSHA. Examples of the latter include furniture manufacturing and the apparel and footwear industry.

Finally, several industries have developed internal guidelines and advisory documents independently of OSHA, such as the semiconductor industry, and the machine tool industry. Some of these various guidelines are program oriented, while others focus more on technical issues of ergonomics, like heights and reaches.

The point is that many of the guidelines that have been developed emphasize managerial programs as much as (or in some cases more than) technical guidance. Furthermore, as of this writing, OSHA has stated its intention to develop or encourage more of these industry-specific guidelines and thus it is prudent for employers to understand and reflect on the nature of these guidelines.

Individual states

Several states in the U.S. have also initiated efforts to promulgate program standards for general industry, all with considerable controversy. The State of Washington prepared a regulation that was to have gone into effect in 2005, but a public referendum in 2003 repealed the plan. The fact that the regulation reached the level of voters in a state-wide referendum is every bit as astounding as the U.S. Congress getting into the act on a Federal level.

As of this writing, the only state with an ergonomics regulation is California. This standard underwent a long, difficult-to-follow series of court cases, but finally remained in effect.

An outline of this standard is provided on the next page. Again, the point here is show the list of organizational activities that are required or encouraged to give you a feel for what types of managerial actions can be taken.

CalOSHA

The State of California has developed standard that has undergone a tortuous history as a result of litigation and appeals. California also requires a written, effective Injury and Illness Prevention Plan (IIPP) applicable for all injuries including musculoskeletal injuries.

Ergonomics: Repetitive Motion Injuries
1997
(Excerpts)

Program

Every employer subject to this section shall establish and implement a program designed to minimize RMIs:

Worksite evaluation — Each job, process, or operation of identical work activity covered by this section or a representative number of such jobs, processes, or operations of identical work activities shall be evaluated for exposures which have caused RMIs.

Control of exposures — Any exposures that have caused RMIs shall, in a timely manner, be corrected or minimized to the extent feasible. The employer shall consider engineering controls, such as workstation redesign, adjustable fixtures or tool redesign, and administrative controls, such as job rotation, work pacing or work breaks.

Training — Employees shall be provided training that includes an explanation of:
- The employer's program
- The exposures which have been associated with RMIs
- The symptoms and consequences of injuries caused by repetitive motion
- The importance of reporting symptoms and injuries
- Methods used by the employer to minimize RMIs.

Scope and application — This section shall apply to a job, process, operation where a repetitive motion injury (RMI) has occurred to more than one employee under the following conditions: Work related; diagnosed by a physician; and the employees incurring the RMIs were performing a job process, or operation of identical work activity. Identical work activity means that the employees were performing the same repetitive motion task, such as but not limited to word processing, assembly or, loading;

Workplace Injury and Illness Prevention Program
1991
(Excerpts)

Management commitment

Your commitment to safety and health shows in every decision you make and every action you take. The person or persons with the authority and responsibility for your program must be identified and given management's full support. You can demonstrate your commitment through your personal concern for employee safety and health and by the priority you place on these issues.

If you want maximum production and quality, you need to control potential work-place hazards. You must commit yourself and your company by building an effective Injury and Illness Prevention Program and integrating it into your entire operation. This commitment must be backed by strong organizational policies, procedures, incentives, and disciplinary actions as necessary to ensure employee compliance with safe and healthful work practices.

Safety communications

Your program must include a system for communicating with employees — in a form readily understandable by all affected employees — on matters relating to occupational safety and health, including provisions designed to encourage employees to inform the employer of hazards at the worksite without fear of reprisal.

Hazard control

An effective hazard control system will identify: hazards that exist or develop in your workplace, how to correct those hazards, and steps you can take to prevent their recurrence.

Accident investigation

Accident investigations should be conducted by trained individuals, and with the primary focus of understanding why the accident or near miss occurred and what actions can be taken to preclude recurrence. It should be in writing and adequately identify the cause(s).

Planning

Planning for safety and health is an important part of every business decision, including purchasing, engineering, and changes in work processes.

Training

As the employer, you must ensure that all employees are knowledgeable about the materials and equipment they are working with, what known hazards are present and how they are controlled. Your supervisors must recognize that they are the primary safety trainers.

"Editorial:"
Should There Be an OSHA Standard?

Ergonomics is good for business and — market forces prevailing — a good case can be made that there is no need for government regulation. As argued in the first part of this book, by systematically applying the principles of ergonomics, employers can cut costs like workers' compensation, turnover, and production inefficiencies. Good ergonomics is good economics.

Yet, standards have their value. Many employers need the authority of a standard to kickstart an activity like ergonomics. A particular case is the government employer, who is often unable to make any changes without a standard. Also, company safety directors and risk managers often find it useful to refer to a standard when there is an internal management disagreement about how to handle an issue. A Federal standard provides a kind of national policy statement that provides direction for employers.

Furthermore, we cannot always rely on the market. To a large extent market forces have failed in regard to ergonomics because of "imperfect information," that is, employers don't know about the costs they are incurring and what is effective in lowering those costs. The market also fails for small employers because their workers' compensation costs are pooled. A large self-insured employer can institute an ergonomics process and effectively cut costs. However, an individual small employer with even the best of internal practices cannot cut insurance costs much, since premiums are set primarily by industry experience.

Moreover, market forces alone may take too long. Employee safety is at stake, and permanent disabilities can be prevented in the meantime.

Finally, there are undoubtedly a number of malfeasant employers who will never invest in worker safety unless forced.

Enforcement Problems

A standard doesn't mean much without having a rational enforcement system in place. Unfortunately, in the U.S., that enforcement system is in shambles. Part of the problem is the practice of regulation generally as it has developed in the U.S. during the past few decades. OSHA itself has proven particularly inept in finding ways to focus on eliminating real hazards while avoiding nitpicking and red tape. I personally have experienced a number of negative encounters with OSHA, including some where the individual inspectors were plainly wrong in what they were advocating.

I personally opposed adoption of the ergonomics standard because it was impossible to enforce. Altho on the one hand, most of the standard (at least as *I* interpreted it) amounted to good management and was consistent with the materials in this book. But on the other hand, if I were an OSHA inspector, I could have cited any employer in the U.S. on any given day using the standard. *That* was the problem.

I should note that there are indeed competent people who work for the agency, including a number of professional ergonomists. But the underlying problem is not one of individuals.

A Different Paradigm

We need a new approach to regulation, one suited for this century, not the past one. Since ergonomics is all about innovation and creativity anyway, why not use the issue of musculoskeletal disorders to see if we could develop a new paradigm?

- We need to find new mechanisms for employers, employees, and OSHA to communicate and make decisions together. OSHA's current approach to industry-specific guidelines is a step in the right direction. The experience of Sweden also provides some guidance along these lines.

- We need to attract senior professionals into the enforcement business. That is, we need to involve experts who have enough experience to be able use judgments rather than look in a book for a paragraph to cite. The inspectors need to wrestle with decisions, just like good managers and good union leaders have to do.*

- We need to insure that the focus is on goals rather than petty details. There needs to be less black and white and less paperwork, while having more emphasis on meeting ultimate objectives.

The premise of ergonomics is that by designing for people we can simultaneously improve human well-being and increase overall efficiency. I suspect this principle applies to Federal regulations and agencies as much as it does to improving assembly tasks and material handling.

---

*In addition to my own experience with and reflections on OSHA's bureaucracy, I have been influenced by Philip Howard's book *The Death of Common Sense* (Warner Books, 1996). After detailing horror stories of misapplied regulations, he concludes that regulators need to be *strengthened* and *allowed to make judgments*, rather than following strict rules like some short-circuited robot.

Other Countries

Canada (British Columbia)

In Canada, the provinces have jurisdiction over workplace safety and health rather than the Federal government. Consequently, regulatory activities vary by province. However, the most developed standard and guidelines are in British Columbia and many labor unions and advocates see these regulations as a model for the rest of Canada.

The standard is promulgated by the Workers Compensation Board (WCB) of British Columbia. Key elements of these provisions are excerpted below. Again, the purpose here is to provide you with a sense of the types of process activities that are encouraged.

Ergonomics (Musculoskeletal Injury – MSI) Requirements
1998
(Excerpts)

Risk assessment

- The employer must identify factors in the workplace that may expose workers to a risk of musculoskeletal injury (MSI).

- When factors that may expose workers to a risk of MSI have been identified, the employer must ensure that the risk to workers is assessed.

- The following factors must be considered, where applicable, in the identification and assessment of the risk of MSI:
  - the physical demands of work activities, including:
    - force required
    - repetition
    - duration
    - work postures
    - local contact stresses
  - aspects of the layout and condition of the workplace or workstation, including:
    - working heights and reaches
    - seating
    - floor surfaces
  - the characteristics of objects handled, including
    - size and shape
    - load condition and weight distribution, and
    - container, tool and equipment handles
  - environmental conditions, including cold
  - the following characteristics of work organization
    - work-recovery cycles
    - task variability
    - work rate

Risk control

- The employer must eliminate or, if that is not practicable, minimize the risk of MSI to workers.

- Personal protective equipment may only be used as a substitute for engineering or administrative controls if it is used in circumstances in which those controls are not practicable.

- The employer must, without delay, implement interim control measures when the introduction of permanent control measures will be delayed.

Education and training

- The employer must ensure that a worker who may be exposed to a risk of MSI is educated in risk identification related to the work, including the recognition of early signs and symptoms of MSIs and their potential health effects.

- The employer must ensure that a worker to be assigned to work which requires specific measures to control the risk of MSI is trained in the use of those measures, including, where applicable, work procedures, mechanical aids and personal protective equipment.

Evaluation

- The employer must monitor the effectiveness of the measures taken to comply with these requirements and ensure they are reviewed at least annually.

Consultation

- The employer must consult with the joint committee or the worker health and safety representative, as applicable, for:
  - risk identification, assessment and control
  - the content and provision of worker education and training
  - the evaluation of the compliance measures taken

- The employer must, when performing a risk assessment, consult with:
  - workers with signs or symptoms of MSI, and
  - a representative sample of the workers who are required to carry out the work being assessed.

Australia

Australia has a standard and two overlapping codes of practice related to workplace ergonomics. Much of the focus is on manual material handling, but a part deals generically with occupational overuse syndrome (i.e., musculoskeletal disorders). The key components, as outlined below, are risk identification, risk assessment and risk control, all in consultation with employees and union representatives.

National Standard for Manual Handling*
1990
(Excerpts)

Risk assessment

An employer shall ensure that manual handling which is likely to be a risk to health and safety is examined and assessed. Risk assessment shall be done in consultation with the employees who are required to carry out the manual handling and their representative(s) on health and safety issues.

The assessment shall take into account the following factors:

- actions and movements;
- workplace and workstation layout;
- working posture and position;
- duration and frequency of manual handling;
- location of loads and distances moved;
- weights and forces;
- characteristics of loads and equipment;
- work organization;
- work environment;
- skills and experience;
- age;
- clothing;
- special needs (temporary or permanent); and
- any other factors considered relevant by the employer, the employees or their representative(s) on health and safety issues.

Risk control

An employer shall ensure, as far as workable, that the risks associated with manual handling are controlled. Risk control shall be done in consultation with the employees who are required to carry out the manual handling and their representative(s) on health and safety issues.

The employer shall, if manual handling has been assessed as a risk:

- redesign the manual handling task to eliminate or control the risk factors; and
- ensure that employees involved in manual handling receive appropriate training, including training in safe manual handling techniques.

Where redesign is not workable, or as a short term or temporary measure, the employer shall:

- provide mechanical aids and/or personal protective equipment, and/or arrange for team lifting in order to reduce the risk; and/or
- ensure that employees receive appropriate training in methods of manual handling appropriate for that manual handling task and/or in the correct use of the mechanical aids and/or personal protective equipment and/or team lifting procedures.

National Code of Practice for Manual Handling
1990
(Excerpts)

Consultation

Consultation [with employees] should occur:

- as early as possible in planning for the introduction of new or modified manual handling tasks, or in the review of existing tasks, to allow for changes arising from the consultation to be incorporated;
- when the employer is identifying the problem areas in order to establish priorities for assessment;
- when determining the approach and methods to be used in assessing the manual handling tasks;
- when decisions are being taken on various control measures to reduce risk factors; and
- when the effectiveness of implemented control measures is being reviewed.

Consultation may occur through formal and/or informal processes, and involve direct and/or representational participation.

Purchasing and design

Purchasing specifications should specify the uses or functions of the plant and equipment, and, where possible, the general performance characteristics required to reduce the risk to health and safety.

Where design or purchase of equipment is planned, the appropriate consultation should occur.

The design of plant, equipment and containers in workplaces needs to provide for a range of physical characteristics of the workforce. Information concerning human dimensions and capabilities should be taken into account to provide an optimum match between the plant or equipment and users.

New and returning employees

Employees newly engaged on a manual handling task or process or returning from an extended absence should, where necessary, be allowed a period of adjustment to build up the skill and ability demanded by the tasks they are required to perform.

Records analysis

Workplace records of injuries should be examined to identify where, and in what jobs, manual handling injuries have occurred.

Direct observation

The direct observation of work areas and of the task being performed will assist in identifying risk. Workplace inspections, audits and walk-through surveys, and the use of checklists can assist in the risk identification process.

The code also includes provisions for training and for workplace design that are not included here. Among other items, it provides a sample checklist, information on good work methods, many examples of lifting aids and equipment to assist in manual handling.

National Code of Practice for the Prevention of Occupational Overuse Syndrome
1994
(Excerpts)

Consultation

Consultation [with employees and union representatives by management] should occur:
- during the design and implementation/purchase of new workplace layout, furniture, work processes and equipment
- when the employer is identifying the areas of risk to establish priorities for assessment
- during the risk assessment process; when determining which risk control strategies (including training) should be applied to prevent or reduce the risk of injuries resulting from tasks involving
  – repetitive or forceful movement or both
  – maintenance of constrained or awkward postures
- when reviewing the effectiveness of implemented control measures

Risk factors

Some of the known risk factors associated with occupational overuse syndrome are:

- awkward body postures;
- poorly designed workstations, equipment, machinery and tools not matched to the employee, including the effects of vibration and sudden impact forces;
- poorly designed tasks, that is, factors such as employee position, forces required and the design and placement of equipment;
- work organization factors which may contribute to demands placed on employees, such as required output, duration and variation of tasks, number and duration of pauses and the urgency of deadlines;
- inappropriate/poor arrangement of job design, for example, the requirement to perform the same repetitive movements; and
- new employees, or those returning to work after an extended absence, being required to perform repetitive movements without a period for adjustment.

Other important factors are the control employees have over the performance of their tasks and in their level of job satisfaction and involvement.

Risk identification

- Analysis of workplace injury and incident records
- Consultation with employees
- Direct observation

Risk assessment

- Working posture
- Duration and frequency of activity
- Force applied
- Work organization
- Skills and experience
- Individual factors

Risk control

- Job design and redesign
- Modify workplace layout
- Modify object or equipment
- Maintenance
- Task-specific (particular) training

Training

Where manual handling has been assessed as a risk, employers should ensure that employees involved in such tasks or jobs receive appropriate training in safe work practices and procedures.

In addition to general training, task-specific (particular) training should be provided to employees. Task-specific (particular) training differs from general

**Training
(continued)**

training in that it is specific to the task, work process, or job. It aims to provide employees with the relevant knowledge and skills to enable them to perform the task in a safe and healthy manner.

Target groups — In addition to the employees involved in manual handling, other target groups requiring training:
- supervisors and managers of employees involved in manual handling tasks;
- employee representatives; and
- employees responsible for the selection and maintenance of plant and equipment, and job and task design and organization.

Training objectives should generally include:
- the prevention and control of manual handling injuries, in particular, those injuries arising from work practices involving repetitive or forceful movement or both, and/or maintenance of constrained or awkward postures;
- the effective implementation of risk identification, assessment and control approaches; and
- the promotion and use of safe work procedures, practices and techniques established for the prevention and control of occupational overuse injuries.

The structure and content of any manual handling training program should be tailored to meet the specific needs and learning requirements of the target group, including the specific needs of employees of non-English speaking backgrounds.

Training review and evaluation — The employer, in conjunction with employees and employee representatives, should regularly review training to ensure training objectives are met.

Training should also be reviewed when there is:
- change in work practices including manual handling control measures;
- change in workplace layouts, task design or organization;
- introduction of new or modified plant or equipment.

Training provided should be commensurate with the associated risks as identified in the assessment process. Training should be provided for all new employees as part of job induction.
- Refresher training should be provided on a regular basis for employees:

**Training
(continued)**

- involved in assessed manual handling tasks to ensure maintenance of safe work practices; and
- returning to the job following extended absence.

Task-specific (particular) training should be provided to employees wherever implementation of the control measures indicates the need. It should be provided by persons skilled and knowledgeable in the specific tasks and job, and in the general approach to manual handling risk control.

Program review

The implementation of this risk control approach, as with any successful systematic process, does not end with the implementation of some change. The effectiveness of the new control measures needs to be reviewed regularly to ensure that the objectives are being achieved and that there are no unforeseen negative outcomes.

Recordkeeping

Records associated with the implementation of [this standard] should be maintained in a central location and be available to employee representatives. Such records will make the tasks of risk identification, and review and evaluation easier.

The records may include information on:
- the prevention program in place to reduce the risk of injury arising from work involving repetitive or forceful movement or both, and/or maintenance of constrained or awkward postures;
- risk identification and assessment;
- design modifications to equipment and work processes;
- risk control measures implemented;
- training and education activities; and
- review and evaluation.

This Code of Practice includes provisions on work organization that are unusual from a North American perspective and are highlighted below.

Work organization

The work should be designed and organized so that the employees are able to regulate their tasks, where workable, to meet work demands. Meeting tight deadlines and peak demands will increase time pressures, reduce control over workflow and may contribute to risk of injury. Bonus and piece-rate systems through their effect on work rate and work organization can be associated with injury.

Bonus and incentive schemes

Some bonus and incentive schemes may contribute to the risk of injury. These schemes may encourage employees to work beyond their individual capacities. Any such scheme should therefore be designed taking these factors into account.

Peak demand

Many jobs have predictable peak periods, which may result in large variations in job demand. The increased risks generated during these peak periods may be prevented by long-term planning of resources and organization of tasks.

Work breaks

Where the job requires a sustained period of repetitive or static (holding or restraining) activity, and it is not possible to provide effective task variation, rest breaks should be provided. The exact length and frequency of such breaks will depend on the nature of the tasks which make up the job.

Working hours

Where work involves repetitive or forceful movement or both, and/or maintenance of constrained or awkward postures, management, supervisors and employees need to be aware of the risk factors associated with extended working hours, for example, overtime, 12-hour shifts, short intervals between shifts and split shifts. Overall organization of shifts will need to be designed to take into account the potential impact on employees of factors such as fatigue and workload.

Displays and control instruments

The appropriate design, selection, arrangement and labeling of displays and control instruments is essential for safe operation of equipment, and will assist in correct posture. A sensible layout of both displays and control instruments will make monitoring easier, reduce the risk of confusion caused by misreading, and reduce visual and postural strain.

These provisions are followed by guidelines on specific applications, such as hand tool design, and workstation layout. They are supplemented by "Guidance Notes" for manual handling in the retail industry, occupational overuse syndrome in keyboard work, and manufacturing industry.

Sweden

Sweden leads the world in workplace ergonomics. The technical innovations that one sees in Swedish workplaces, the educational materials and programs, the research on the work life, and the genuine interest which many Swedish people take in the work environment are all remarkable.

The legal and regulatory provisions, as outlined below, are also probably unique in the world. The scope of these regulations is astonishingly broad, comprising "all physical, psychological and social conditions of importance for the work environment."

In 1978, Swedish parliament enacted a law that fundamentally changed the conceptual basis of workplace safety. The term "work environment" is noteworthy and signifies a concern for issues that are broader than those of traditional "occupational health and safety." In 2001, this law was expanded and included a change in the name of the agency responsible for workplace safety. Again, the name change is significant, from "Labor Inspectorate" and "National Board of Occupational Safety" to the "Work Environment Authority."

The law is supplemented by two broad regulations that provide the full regulatory context for prevention of MSDs. Key elements of these provisions are outlined below.

Work Environment Act

1978, 2001
(Excerpted)

Employee representatives

Safety officers — At every location of employment where five or more employees are regularly engaged, one or more of the employees shall be appointed safety officer [by their union].

Safety committees — At a workplace where fifty or more persons are regularly employed, there shall be a safety committee consisting of representatives of the employer and of the employees [that is, the union].

Employee participation

The safety officer shall participate in the planning of new premises, equipment, work processes, work methods or alterations to existing ones.

The safety committee shall participate in the planning of work with respect to the work environment and follow up the implementation of that work. It shall carefully monitor developments with respect to issues relating to protection against illness and accidents and is to promote satisfactory work environment conditions.

Training — Employers and employees shall be jointly responsible for ensuring that safety officers receive the requisite training.

Systematic Work Environment Management
2001
(Excerpted and edited)

Scope

All questions with a bearing on the working environment —There are many different factors at work by which the employee is physically and mentally affected. Together these factors make up the employee's total working environment. They include, for example, noise, air quality, chemical health hazards and machinery, as well as organizational conditions such as work load, working hours, leadership, social contacts, variation and the possibility of "rest and recovery."

The employer needs to take into account all factors potentially impacting on the individual persons' work situation. This does not only mean things capable of adversely affecting health and safety. A good work environment contributes toward good health and means more than the absence of illness and accidents.

Management

Systematic work environment management shall be included as a natural part of day-to-day activities. It shall comprise all physical, psychological and social conditions of importance for the work environment.

The employer shall give the employees, safety delegates and pupil safety delegates the possibility of participating in systematic work environment management.

There shall be a work environment policy describing how working conditions in the employer's activity shall be in order for ill-health and accidents at work to be prevented and a satisfactory work environment achieved.

There shall be routines describing how systematic work environment management shall proceed.

The work environment policy and the routines shall be documented in writing if there are at least ten persons employed in the activity.

Accountability

The employer shall assign responsibility to managers, supervisors or other employees to working for the prevention of risks at work and the achievement of a satisfactory working environment. The employer shall see to it that the persons allotted these tasks are sufficient in number and have the authority and the resources that are needed. The employer shall see to it that those allotted the tasks have sufficient competence for the conduct of a well-functioning systematic work environment management.

Knowledge

The employer shall also see to it that those responsible have sufficient knowledge of:

- rules material to the working environment,
- physical, psychological and social conditions implying risks of accidents and ill-health,
- measures to prevent ill-health and accidents, and
- working conditions conducive to a satisfactory working environment

Risk assessment

The employer shall regularly investigate working conditions and assess the risks of any person being affected by ill-health or accidents at work.

Ergonomics for the Prevention of Musculoskeletal Disorders
1998
(Excerpts)

Responsibilities of the employer

Work postures and movements — The employer shall as far as is practically possible design and arrange work and workstations in such a way that the employees can use work postures and working movements which are favorable to the body.

Manual handling and other exertion of force — The employer shall ensure that work requiring the exertion of force is, as far as is practically possible, designed and arranged in such a way that the employee can work with a work object, work equipment, controls, material or transfer of persons without being exposed to physical loads which are dangerous to health or unnecessarily fatiguing.

Physically monotonous, repetitive, closely controlled or restricted work — The employer shall ensure that work which is physically monotonous, repetitive, closely controlled or restricted does not normally occur. If special circumstances require an employee to do such work, the risks of ill-health or accidents resulting from physical loads which are dangerous to health or unnecessarily fatiguing shall be averted by job rotation, job diversification, breaks or other measures which can augment the variation at work.

Job decision latitude — The employer shall ensure that the employee has such opportunities of influencing the arrangement and performance of his own work that sufficient variation of movement and recuperation can be achieved.

Knowledge, skills, and information — The employer shall ensure that the employee has sufficient knowledge concerning:

- suitable work postures and working movements,
- the proper use of technical equipment and aids,
- the risks entailed by unsuitable work postures, working movements, and unsuitable manual handling
- early indications of overloading of joints and muscles

Non-employer responsibilities

Employees — The employee shall be attentive to the employer's instructions concerning the avoidance of physical loads which are dangerous to health or unnecessarily fatiguing. An employee judging that a task can entail such loads shall notify the employer accordingly.

Manufacturers, importers, and suppliers — [These providers] shall as far as is practically possible ensure that the technical devices, substances and packagings delivered do not cause physical loads which are dangerous to health or unnecessarily fatiguing in connection with installation, normal use, maintenance or other commonly occurring handling. Where necessary, the delivery shall be accompanied by information as to how the devices, substances and packagings are to be handled in an ergonomically safe manner.

Construction projects — Whoever commissions building or civil engineering work shall at the project preparation stage, as far as is practically possible, prevent physical loads which are dangerous to health or unnecessarily fatiguing occurring during the construction phase or in the intended use of the building or structure. This applies particularly to selection of materials, accessibility and transports.

(The regulation continues with additional background, interpretations, checklists and related guidance.)

Interpretations and Enforcement*

The stipulations quoted in the previous pages provide the legal context in Sweden for workplace ergonomics and other safety and health issues. However, it is noteworthy that the primary mechanism for interpretation and enforcement of the requirements in specific industries and worksites does *not* lie with the government. Rather, the core of the enforcement system lies with joint negotiation and cooperation between trade associations and organized labor.

Strong and Efficient Unions

The Swedish labor movement is one of strongest in the world, with well over 90 percent of the blue-collar workforce unionized. Over 70 percent of Swedish white-collar workers and about 70 percent of supervisors are unionized.

Moreover, the unions use power responsibly, as Swedish business leaders often agree. In particular, Swedish unions have traditionally urged and supported the modernization of industry, the use of advanced technology, and the improvement of efficiency at work. Swedish union representatives are well educated, stemming from a large network of union schools and study circles (of which there are few equivalents in the U.S.). The unions themselves are efficiently run and well-organized by industry group.

Powerful Trade Associations

Employers are also well organized in Sweden, in part as a response to the labor movement. Nearly all private firms belong to the Swedish Federation of Employers, a tightly organized group with the ability to impose decisions on its members, either through its bylaws or through the union contracts it negotiates.

The Federation of Employers meets regularly with the Federation of Labor to negotiate a central contract that sets pay increases and other key matters that cover the bulk of the Swedish workforce. Subsequent negotiations on an industry level, and then in individual workplaces, implement these central agreements plus provide flexibility to adapt to local conditions.

For example, unions and employers negotiated procedures on how to implement the Swedish Work Environment Act. Similarly, joint labor-management groups negotiated methods to conduct systematic health and safety education and to develop comprehensive training materials. In the United States, no equivalent structure exists where such

---

*I now resume my comments based on living and working in Sweden; these are not excerpts from another source.

nationwide agreements between management and labor can be reached.

This structure provides an unusual advantage over the U.S. and other countries around the world. Swedish employers and unions are able to discuss and resolve problems that in the United States are beyond anyone's control. There is a system that enables enforcement of a national policy at the local level, but with flexibility to accommodate the special needs of each individual workplace.

It should be emphasized that most of this activity occurs privately between employers and unions without government involvement. In the U.S., without this structure, government is much more often involved, typically by federal policymakers who attempt to develop generic regulations that fit every industry, workplace, and circumstance . . . an impossible task.

Implications

The experience of Sweden can be helpful on two levels. The first is that the list of employee rights and management duties provides insights on the process used by employers in the country that is most advanced in ergonomics.

The other level has to do with how to develop, implement, and enforce guidelines in a rationale way. To be sure, most countries cannot just decide to set up an efficient labor relations system along the lines developed in Sweden. However, regulatory agencies can and should learn from this example and adapt the concepts as best as possible into their own systems.

Additional Countries

New Zealand

The New Zealand Department of Labour, in cooperation with the Accident Compensation Corporation, has published guidelines for the prevention and management of Occupational Overuse Syndrome. This document reads more like a training booklet than a set of regulations. The contents are generally equivalent to the standards and guidelines outlined above, with the exception of having an additional section on rehabilitation and the role of occupational therapists and physiotherapists.

Singapore

Singapore, with its advanced economy and high standard of living, has developed both a *Code of Practice on Manual Handling* and *Guidelines on Office Ergonomics*. A broader set of ergonomics guidelines has also been discussed.

Spain

The Spanish government has published a considerable number of guidelines regarding various issues, including psychosocial issues and cognitive ergonomics. Most of these are technical, rather than addressing organizational issues in setting up workplace programs. As an unintended benefit for North Americans, these guidelines are good general sources of information on ergonomics in Spanish.

United Kingdom

The focus in Britain at the time of this writing is on manual material handling. The Manual Handling Operations Regulations 1992, as amended in 2002, apply to a wide range of manual handling activities, including lifting, lowering, pushing, pulling or carrying. The load may be either inanimate, such as a box or a "trolley" (a "cart" in American parlance), or animate — a person or an animal.

The regulations require employers to: *avoid* the need for hazardous manual handling, so far as is reasonably practicable; *assess* the risk of injury from any hazardous manual handling that can't be avoided; and *reduce* the risk of injury from hazardous manual handling, so far as is reasonably practicable.

These regulations were promulgated in the context of other rules that require employers to consult and keep safety representatives and employees up to date (Safety Representatives and Safety Committees Regulations 1977 and the Health and Safety Consultation with Employees Regulations 1996).

Part III — Program Tools

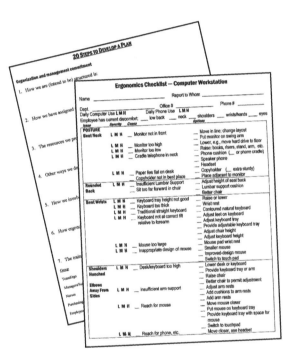

Chapter 10
Worksheets

This chapter contains a variety of worksheets and survey tools for use in your program.

Overview

Program development — The first group of worksheets addresses program issues — where to get started, questions to be answered to develop a written plan, a quarterly planning guide, and an audit. Some of these are brainstorming exercises for TeamErgo that you would use only once. Others are tools that are intended for ongoing use.

Task analysis worksheets — Several versions of checklists and worksheets are provided. Some serve different functions, others provide different versions of the same type of worksheet to more easily fit your organizational style.

Employee feedback forms — Also included is a simple discomfort survey form along with guidance for its use, plus several sample employee feedback forms.

Innovation, quality, and efficiency

These worksheets focus on the prevention of MSDs. But as we have seen, ergonomics can be an effective tool for promoting innovation and improving efficiency. As you work thru these worksheets, keep in mind these broader issues. You often still look for the same things: awkward postures, wasted motions, etc. But your focus changes slightly from "Would this cause an MSD?" to "Is this a source of inefficiency or waste?"

In actuality there is a huge overlap in the answers to these questions, so you tend to address both simultaneously, especially if you consciously try to do so.

Scoring methods

No task scoring systems are included here. A good system in the public domain and available on the internet is Rapid Upper Limb Assessment (RULA).

Copyrighted materials

The checklists and worksheets are available electronically for your use (see Table of Contents). However, they are copyrighted. You have permission to photocopy these materials for your internal use at your own facility, but you may not sell them or distribute them in any way beyond your own internal use at a single workplace. Further licensing may be possible by contacting the author at dan@danmacleod.com.

Brainstorming Activities

Topic: Priority jobs for improvement

Based on your current knowledge, list the priority areas in your facility that an ergonomics program should address:

1.

2.

3.

4.

5.

Topic: Discovering priorities

What additional information do we need to identify priority areas?

1.

2.

3.

4.

5.

Topic: Gaining management commitment

The term "management commitment" is used frequently, but can be somewhat vague. What does "management commitment" really mean as related to ergonomics?

A. List five examples of concrete actions or activities that show management commitment to ergonomics. (Pretend you are a government inspector investigating an employer. What evidence would convince you that management is committed to ergonomics?)

1.

2.

3.

4.

5.

B. How should TeamErgo go about getting management commitment for the ergonomics program at your facility? List five ways of obtaining this commitment.

1.

2.

3.

4.

5.

Topic: Gaining Management Commitment

This is a quick exercise to help people understand the financial costs associated with poor ergonomics:

1. List the types of adjustment that are available in cars to accommodate different sizes of people:

2. How do you feel when you drive a car that is not adjusted for you?

3. What would happen if you drove eight or 10 hours per day, day after day, in a position that was not right for you?

4. What would you do if you went to a car dealer to buy a new car and the seat was not adjustable, not even forward and backward?

5. What happens when employees work in awkward and uncomfortable positions?

Topic: Dovetailing with other initiatives

This is a very important exercise.

	What else is going on in your facility?	How does the ergonomics process support and take advantage of that activity?
Examples:		
Organizational Mission Statement(s)		
Major Organizational Goals		
Training		
Quality Improvement		
Efficiency		
Empowerment (or teams)		
Cultural Change		
Employee Relations		
Facility Renovations or Expansions		
Other		

List five ways in which you will deliberately structure the ergonomics process in your facility to fit into these activities.

1.

2.

3.

4.

5.

How will you identify these issues in the presentations and training that you will conduct?

Topic: Ensuring employee participation

What type of employee participation would you want to see in an ergonomics program? List three examples:

1.

2.

3.

How could you ensure this participation?

1.

2.

3.

Topic: Early recognition of employees with musculoskeletal disorders (MSDs)

Current medical advice is that the sooner a person with an emerging MSD is recognized, the easier (and cheaper) it is to treat.   To accomplish this involves a change in our traditional philosophies ("Why have a pain checked out; maybe it'll go away," or "I'll wait till it's a real problem before I go see the doctor.") as well as our organizational systems.

What happens now in your facility when employees experience symptoms of MSDs?

What tangible steps could be taken in your organization to (a) encourage early reporting and (b) respond appropriately? (List three; there are potentially many more.)

1.

2.

3.

What obstacles will you need to overcome to initiate or improve these steps in your organization? (List three.)

1.

2.

3.

Topic: Communications

This is an especially important worksheet.

List five mechanisms that exist in your facility that are used for communications. (Example: cafeteria bulletin board)

1.

2.

3.

4.

5.

For each of these, how exactly can you use them to promote ergonomics?

	Particular strength of this mechanism	Types of messages that fit this mechanism best	Person responsible	Deadline
Example: Bulletin Board	Everyone goes by at least once/wk.; easy to change	Before-and-after photos	Pat	18th
1.				
2.				
3.				
4.				
5.				

(Good communications isn't hard; you just need to plan for it.)

20 Steps to Develop a Plan

This list is to help you put on paper things you are doing or planning. You don't need to address every question.

Organization and management commitment

1. How we are (intend to be) structured is:

2. How we have assigned (will assign) responsibility is:

3. The resources we provide (will provide, etc.) to the ergonomics effort are:

4. Other ways we demonstrate commitment are:

5. How we involve employees in our ergonomics process is:

6. How ergonomics dovetails with other initiatives and activities in our facility is:

7. The training that site personnel have received (will receive) is:

Group	Topics	Amount of Time
TeamErgo		
Managers/Supervisors		
Nurses		
Purchasing/Facilities		
Employees		

Communication

8. The approaches we take to inform the entire facility of our activities are:

9. We keep abreast of developments in other facilities by:

10. The steps we take to encourage early reporting of MSDs are:

Identifying issues and improvements

11. The way we respond to reports of MSDs and MSD risk factors is:

12. The key records we analyze to identify ergonomic problems are:

13. Other steps we take to identify problems and set priorities are:

14. The process we use to evaluate tasks and solve problems can best be described as:

Improvements

15. The process we use to set timetables to make changes and track progress is:

16. Administrative programs we use to reduce CTDs are:

Medical management

17. We provide medical care services to our employees with CTDs in the following way:

18. We involve medical providers in our ergonomics program by:

19. The system we use to provide work for employees with restrictions is:

Measuring progress

20. The methods we have chosen to measure our progress are:

Summary

Where we have been to date is:

The two or three most important things we have to do (or barriers we have to overcome) are:

The two or three major unanswered questions on implementing our ergonomics program are:

The two or three immediate steps we need to take are:

Ergonomics Quarterly Planning Guide
and Progress Report

Location: _____ Quarter _____ Year _____

> Directions: At the beginning of each quarter, list the activities you intend to do. At the end of the quarter, list what you accomplished, and develop your goals for the next quarter. If you do not meet a particular goal (which is expected in some cases), simply describe the reason why not.

1. Overall program
Based on a review of your written program and overall implementation plan, what are your plans this quarter to address your overall program development?

Examples:
Conduct employee surveys, initiate supervisor training, review injury/illness records.

Goals	Accomplishments
	(to be completed at end of quarter)

2. Training
What training sessions/safety talks will be conducted this quarter? (Who and how many people?)

Goals	Accomplishments

3. Workstations evaluations (using Ergonomics Worksheet)
Based on your priorities, what work areas will a team evaluate this quarter?

Goals	Accomplishments

4. Actions
Based on the above evaluations, what improvements are planned?

Goals	**Accomplishments**
(Items may need to be added during committee meetings as the quarter progresses)	

5. Major Project(s)
For ongoing major projects that will take longer than one quarter to address, what steps will be taken <u>this quarter</u>?

Goals	**Accomplishments**

6. Other
Are there any other plans? (*Or at end of quarter:* Did you have any other accomplishments?)

Goals	**Accomplishments**

7. Things that work
Of the things that you worked on this past quarter, did anything go particularly well? What was the greatest insight you gained?

_____

_____

8. Obstacles
Were there any particular obstacles?

_____

_____

Ergonomics Program Audit

The following is a program audit developed to help managers develop and maintain good ergonomics programs. It was designed to follow the framework used in Chapter 5 — *Setting Up an Ergonomics Program*:

1. **Organization**
2. **Training**
3. **Communication**
4. **Job Analysis**
5. **Job Improvements**
6. **Medical Management**
7. **Monitoring Progress**

This audit is intended to be used primarily as a self-audit to reveal gaps in plant programs and to identify helpful ideas to further program development. It can also be used as a training tool and working document for ergonomics committees and coordinators. In some ways, this audit can also be seen as a summary of Chapter 5, *Setting Up an Ergonomics Program.*

This audit form is not an interpretation of any other guidelines or standards. The items listed are not requirements nor are they appropriate for all workplaces. In particular, this audit was designed for large organizations with considerable problems and may be of limited use for small organizations (unless adapted appropriately). The amount of activity for each line item should be in proportion to the magnitude of ergonomic issues in each workplace. The points assigned are somewhat arbitrary, but do reflect some degree of importance.

Program Audit

1. Organization	Points		Comments
	Possible	Earned	
A. Management Responsibility			
1. Does upper management:			
a. Clearly assign responsibility for program elements?	1		
b. Monitor progress on ergonomics projects, program implementation, etc.?	1		
c. Review ergonomics issues in staff meetings?	1		
1. Are ergonomics issues included in performance reviews for all managers?	1		
2. Does this facility have an ergonomics coordinator?	1		
3. Does the coordinator meet at least monthly with top management?	1		
4. Has an ergonomics budget been established for this facility?	1		
5. Has a special ergonomics maintenance crew (or person) been assigned?	1		
B. TeamErgo			
1. Does TeamErgo:			
a. meet regularly to coordinate functioning of all program elements?	1		
b. review and update program goals and objectives at least quarterly and set priorities for action?	1		

Award partial points (e.g., ½) when appropriate.

Organization (continued)	Points		Comments
	Possible	Earned	
c. provide systematic and organized follow-up on implementing ergonomic solutions?	1		
2. Are minutes kept to summarize discussions, document decisions made, and define action items?	1		
C. Employee Involvement			
1. Are hourly employees or union representatives actively involved in the program?	1		
2. Are employee suggestions sought and used in the program?	1		
3. Are employee meetings held to get input before plant floor changes are made?	1		
4. Are employees provided opportunity for input when job evaluations are made?	1		
D. Written Program			
1. Are top managers and staff familiar with the written program?	1		
2. Is the site's written program reviewed whenever significant changes are made in personnel, operations, or plant layout; or at least annually?	1		
3. Are specific goals outlined in an annual operations plan?	1		
4. Is the ergonomics program well-integrated with other activities?	1		
Total Points for Organization	**20**		

2. Training	Points		Comments
	Possible	**Earned**	
1. Have all superintendents, managers, supervisors and engineers received overview training on:	2		
a. the ergonomics program and activities?			
b. basic ergonomic principles?			
c. musculoskeletal disorders?			
d. the need for early medical intervention?			
e. the importance of adhering to employee medical restrictions?			
2. Additionally, have all TeamErgo members received training on:	2		
a. program guidelines?			
b. general ergonomics?			
c. musculoskeletal disorders?			
d. job evaluations/problem solving?			
3. Have all medical providers been trained in the medical management program?	2		
4. Have maintenance personnel been trained in ergonomics?	1		
5. Are all employees and new hires:	2		
a. provided overview ergonomics training?			
b. told of the necessity of reporting symptoms?			

Training (continued)	Points		Comments
	Possible	**Earned**	
c. shown how to properly use tools and equipment?			
d. shown the best ergonomic methods to perform specific tasks?			
6. Is there follow-thru to check on the work skills of new hires and reinstruct as necessary?	1		
Total Points for Training	**10**		

3. Communications

	Possible	Earned	Comments
1. Are updates on ergonomics activities provided to employees thru bulletin boards or newsletters?	2		
2. Are updates on activities regularly supplied to meetings of managers, supervisors, employees, etc.?	3		
3. Is feedback on the status of ideas and suggestions given to affected employees and supervisors?	2		
4. Is credit given to project teams and originators of ideas?	1		
5. Are minutes of TeamErgo meetings distributed to key people?	1		
6. Do plant representatives attend all appropriate conferences (corporate meetings, trade association, meetings, etc.)?	1		
Total Points for Communications	**10**		

4. Job Analysis	Points		Comments
	Possible	Earned	

1. Are OSHA 300 logs and other records analyzed periodically for:
 	
  a. MSDs by job and type?
 	
  b. rates calculated?
 	
  c. graphs plotted to compare jobs or departments?
 	
  d. High-injury jobs targeted?
	2		

2. Are employee discomfort surveys conducted in targeted areas and results analyzed and used?	1		
3. Are *Ergonomics Innovation Worksheets* (or equivalent) used and kept on record?	1		
4. Are videotapes made of jobs in conjunction with these checklists and reviewed in brainstorming sessions?	2		
5. Are plant or area surveys conducted at least monthly to identify issues?	1		
6. Whenever plant or equipment changes are planned, is ergonomics made part of the planning process (employees and committee members involved, *Ergonomics Innovation Worksheets* used, other plant coordinators contacted, etc.)?	2		
7. Is experience with turnover and unpopular jobs evaluated to help set priorities for job improvements?	1		
Total Points for Job Analysis	**10**		

5. Job Improvements	Points		Comments
	Possible	Earned	
1. Is there at least one major ergonomics project in active progress?	3		
2. Are easy-to-implement changes made, even though not necessarily high-priority issues?	3		
3. Are suggestions for improvements implemented from:			
a. completed *Ergonomics Innovation Worksheets*?	2		
b. employee meetings?	2		
c. supervisor meetings?	2		
d. managers, superintendents, and engineers?	2		
4. Are all job improvements (and unsuccessful attempts) entered on the Ergonomics Log?	1		
5. Have administrative changes been made to reduce CTD risk factors (less overtime, job rotation, etc.)	1		
6. Are costs of turnover, absenteeism, CTDs, poor quality, etc. included to justify investment in plant and equipment changes?	1		
7. Are developments in your industry and general industry being monitored for useful ideas?	1		
8. Are brainstorming sessions regularly held (watching videos of tasks)?	2		
Total Points for Job Improvements	**20**		

6. Medical Management	Points		Comments
	Possible	Earned	
1. Are designated health care providers available at this facility?	1		
2. Do medical providers perform regular workplace walkthrus?	1		
3. Are employees in targeted areas provided physical examinations for MSD symptoms?	1		
4. Are employees encouraged to report symptoms of MSDs?	2		
5. When an employee reports MSD symptoms does:			
a. the medical provider conduct a physical examination?	3		
b. the medical provider treat the employees in accordance with the treatment algorithms for upper extremity complaints?	3		
6. When an employee is placed on restricted duty, is the job reviewed to assure appropriateness?	2		
7. Are exercise programs used for warm-up and stretch breaks?	1		
8. Are work-hardening practices followed in returning workers from restricted duty?	1		
9. Are new hires and those returning from injury allowed break-in periods to recondition themselves for job demands?	1		
10. Is the OSHA 200 Log accurate and up-to-date?	4		
Total Points for Medical Management	20		

7. Monitoring Progress	Points		Comments
	Possible	Earned	
1. Can you document improvements in any of the following: a. reduced MSD rates? b. reduced turnover? c. reduced absenteeism? d. increased efficiency? e. increased quality? f. increased morale? g. employee discomfort surveys?	2		
2. Are MSD rates graphed quarterly to compare trends?	2		
3. Are ergonomics logs of improvements summarized and communicated to plant personnel?	2		
4. Have slides, photos, or videotapes been made of before-and-after changes?	2		
5. Have any quantitative studies been made of before-and-after changes?	2		
Total Points for Monitoring Progress	10		
Grand Total Points	100		

Task Analysis

Introduction

"Analysis" literally means to take a whole and break it into its subparts (it's the opposite of "synthesis"). Task analysis then means simply to break a task down into its subparts.

The reason for breaking down a task into subparts is that often, if you look at the task as a whole, it is too much to comprehend all at once, and you end up not gaining much insight. However, if you focus just on one thing for a while, move onto another, then you start to really see what is happening.

For our purposes, task analysis means to break a job down into:

- Steps of the job — "reach for part A, insert into part B, etc."
- Parts of the body — wrists, arms, back, etc.
- Ergonomics issues — force, awkward postures, etc.

Then, for each ergonomic issue affecting a specific body part during a particular step of the job, we ask ourselves, "Can we think of an easier way to do this by applying the basic principles of ergonomics?"

Finally, after looking at the task in question in this focused fashion, part by part, then we take a step back and look at the whole job again and ask, "What have we learned? What are our overall insights and conclusions?"

To assist you in conducting task analysis, this part of the *Ergonomics Kit* contains a number of different types of worksheets, each with its own purpose:

Ergonomics innovation worksheet

This is TeamErgo's primary problem-solving tool. The word "innovation" is a critical one, since the goal for these evaluations is to figure out creative ways to make improvements.

The regular version of this worksheet is followed by a longer training version that you may wish to use as you get started. It is more comprehensive to build your confidence, but it is too cumbersome to use on a continuing basis.

A third, self-assessment version is designed for employees themselves to use to review their own work areas.

Walkthru survey

If you wish to use a worksheet when performing walkthru evaluations, this is a basic version.

Computer workstation checklist

A special checklist for computer workstations is also provided. The focus is not on brainstorming and innovation as with the above worksheets, rather it is on identifying which employees need what types of work aids or adjustments.

Task Analysis Schematic
To separate a whole into its subparts for individual study.

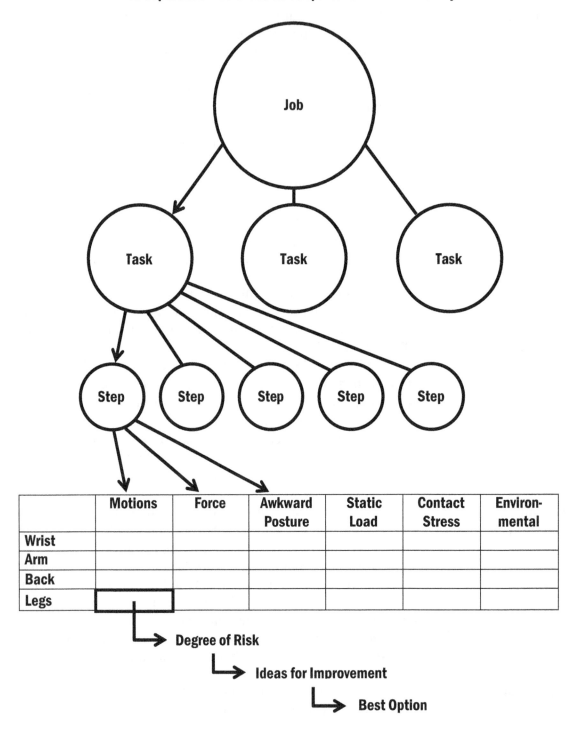

	Motions	Force	Awkward Posture	Static Load	Contact Stress	Environ-mental
Wrist						
Arm						
Back						
Legs						

→ Degree of Risk

↳ Ideas for Improvement

↳ Best Option

Comments:
- Ideally, an ergonomics analysis is performed for each step. However, sometimes evaluating the task as a whole is sufficient.
- The degree of risk can be quantified or can be categorized as "low," "medium," or "high." Even if a task is rated "low" or "medium," if you can think of a way to make it better at a reasonable cost, do it.

Ergonomics Innovation Worksheet

(Make sure to explain your purpose to the employees at the task you are evaluating.)

Area _____ Date of Evaluation _____

Task _____ Shift 1 ____ 2 ____ 3 ____

Steps of the task

_____   _____   _____

_____   _____   _____

_____   _____   _____

Ergonomics Issues   **Ideas for Improvement or Comments**

1. Awkward postures?

bent wrists

elbows from body

bent/twisted back

bent neck

2. Excessive forces?

grasping or pinching forces

push/pull arm forces

loads on back

3. Anything not in easy reach?

reach envelope:

- full arm

- fore arm

4. Not at right height?

over shoulders/below knees

elbow height

equipment height relationships

5. Excessive motions?

hands

arms

back

Ergonomics Issues	Ideas for Improvement or Comments

6. Unnecessarily fatiguing?

static loads:

- grip

- arm

7. Pressure points?

tool grip

hard edges/surfaces

hard floor

8. Poor clearance and access?

bump/not fit

can't see

9. Freedom to move & stretch?

constant sitting

stand in one place

10. Uncomfortable environment?

vibration

temperature extremes

glare, shadows, too bright or dark

Additional Information

What suggestions and feedback do employees/supervisors have?

Have there been injuries reported from this task?
- If so, how many?
- What have you learned from reading Injury/Illness Report
- What have you learned from additional discussions with the person(s) reporting the injury?

Where do employees typically experience the most discomfort, if any?

hands/ wrists	__ none	__ low	__ medium	__ high
elbows/forearms	__ none	__ low	__ medium	__ high
shoulders	__ none	__ low	__ medium	__ high
neck	__ none	__ low	__ medium	__ high
back	__ none	__ low	__ medium	__ high
legs	__ none	__ low	__ medium	__ high
feet	__ none	__ low	__ medium	__ high

What are the top 2-3 issues for this task?

Use the attached sheets to help brainstorm possible improvements. Make as many copies as needed.

Issue _____

Options for Improvement
(List as many as you can. Include "harebrained" ideas to stimulate creativity.)

_____   _____

_____   _____

_____   _____

_____   _____

_____   _____

Action Items:

Short Term   **Long Term**

_____   _____

_____   _____

_____   _____

Sketch

What can you "trystorm" right now?
(that is, what can you mock up to test an idea)

Innovation Potential

For each of the issues you identified above, use the forms on the following pages to brainstorm ideas and list as many options for improvement that you can think of. Use the following questions to stimulate your creativity.

Probing questions

What improved types of tools are possible?

What types of mechanical assists might be used?

What changes in layout would help?

What changes in product or packaging design would help?

Would improvements in the overall material handling system help?

Would changes in the overall work process help?

Is there a completely different way of doing the job?

Mindset questions

If you were a Yankee inventor living in 1820
and had no electricity or power, how would you
do this job?

If you had unlimited resources, what would you do?

If there is an automatic way of doing this task
that is too expensive or not feasible in this
case, can you think of some half-way
modification that *is* feasible?

Have you ever seen a different way of doing
this task? What implications does that have in
this case?

Does a similar task exist in another industry?
How do they do it there? What are the
implications for you?

Ergonomics Innovation Worksheet
(Training Version)

This form is mostly self-explanatory. However, note the following:

Training version — The long version is useful when you first start, to lead you thru the process and help make sure you don't miss things. But once you get the hang of it, the long version is cumbersome, and the previous shorter one is easier to use.

Explain why you're there — Courtesy is important, so before you do anything, introduce yourself. Explain your mission, for example, by saying you are trying to find safer, easier ways of doing the task. Sometimes it is helpful to say specifically, "I'm evaluating the task, not you." As appropriate, indicate that you may be asking questions about the job, any sources of discomfort, and any suggestions for improvement.

Steps of the task — Before you start problem-solving, make sure you understand what the person you are watching is doing — step by step. You do not need to be elaborate on this, but do your best to define the various steps of the task.

Low-medium-high — Don't make a big deal about the differences between low-medium-high. That is not the goal of the *Innovation Worksheet*. If you need to make precise determinations about risk, there are better approaches to use. The goal here is to promote thinking about improvements. The purpose of asking you to rate each issue is primarily to force you to reflect on potential problems one at a time.

Using the Self Assessment Form

Following is another short form in a different framework and from the perspective of the employee. Both forms cover the same territory and it is mostly a matter of your choice as to which to use.

Ergonomics Innovation Worksheet

Department _____ Date _____

Task _____ Shift 1st ____ 2nd ____ 3rd _____

Steps of the task

_____   _____   _____

_____   _____   _____

_____   _____   _____

Ergonomics Issues

Directions: This column serves as a "mind-jogger" and a systematic method to help you focus and identify issues. The purpose is not to document precise job conditions or make firm distinctions between low, medium, and high. The purpose is to help you reflect on all the issues and to prompt you to think about ways to make improvements.

Ideas for Improvement or Comments

Directions: This side of the page is the more important side. List as many things as you can think of to make improvements. Even if you check an item as "low" you may still think of a low-cost improvement that makes sense. But especially if you think some issue is a "yes", "high", or "extreme", list ideas. Don't worry about feasibility (yet); just write down ideas.

Back

Repetitive back motions	__ low	__ medium	__ high

Loads or forces on back

lifting	__ low	__ medium	__ high
carrying	__ low	__ medium	__ high
pushing	__ low	__ medium	__ high
pulling	__ low	__ medium	__ high

Awkward postures

bending	__ low	__ medium	__ extreme*
bent neck	__ low	__ medium	__ extreme
twisting	__ low	__ medium	__ extreme
			(* or constant)
Whole body vibration	__ low	__ medium	__ high

Arm

Left

Push/pull force	__ light	__ medium	__ heavy
Static load on shoulder	__ none	__ occasional	__ constant
Arm motions	__ low	__ medium	__ high
Elbow away from body	__ neutral	__ medium	__ extreme
Pressure point (forearm)	__ light	__ medium	__ heavy

Right

Push/pull force	__ light	__ medium	__ heavy
Static load on shoulder	__ none	__ occasional	__ constant
Arm motions	__ low	__ medium	__ high
Elbow away from body	__ neutral	__ medium	__ extreme
Pressure point (forearm)	__ light	__ medium	__ heavy

Sample ideas
__raise load off floor
__lower load height
__change load shape
__add hand holds
__use scissors lift
__use power tilter
__change layout
__tilt the surface
__mechanical arm
__vacuum hoist
__reduce weight
__add a person
__use cart
__use conveyor

Sample ideas
__change layout
__improve heights
__smaller surface
__cutout
__fixture the part
__fixture the tool
__counter-balancer
__mechanical assist
__tilt work surface
__arm rest
__angle tool handle
__pad hard edges

| **Ergonomics Issues** | | | | **Ideas for Improvement or Comments** |

Hand/Wrist

<table>
<tr><td colspan="4"></td><td>*Sample ideas*</td></tr>
<tr><td colspan="4">Left</td><td>__power tool</td></tr>
<tr><td>Grasping force</td><td>__ light</td><td>__ medium</td><td>__ heavy</td><td>__different tool</td></tr>
<tr><td>Static grip</td><td>__ none</td><td>__ occasional</td><td>__ constant</td><td>__improved tool</td></tr>
<tr><td>Wrist motions</td><td>__ low</td><td>__ medium</td><td>__ high</td><td>__tilted work surface</td></tr>
<tr><td>Bent wrist</td><td>__ neutral</td><td>__ medium</td><td>__ severe</td><td>__angled tool grip</td></tr>
<tr><td>Pressure on palm</td><td>__ light</td><td>__ medium</td><td>__ heavy</td><td>__improved tool grip
__grip wrap</td></tr>
<tr><td>Vibration</td><td>__ none</td><td>__ low</td><td>__ heavy</td><td>__fixture for part
__fixture for tool</td></tr>
<tr><td colspan="4"></td><td>__better gloves</td></tr>
<tr><td colspan="4">Right</td><td>__dampen vibration</td></tr>
<tr><td>Grasping force</td><td>__ light</td><td>__ medium</td><td>__ heavy</td><td></td></tr>
<tr><td>Static grip</td><td>__ none</td><td>__ occasional</td><td>__ constant</td><td></td></tr>
<tr><td>Wrist motions</td><td>__ low</td><td>__ medium</td><td>__ high</td><td></td></tr>
<tr><td>Bent wrist</td><td>__ neutral</td><td>__ medium</td><td>__ severe</td><td></td></tr>
<tr><td>Pressure on palm</td><td>__ light</td><td>__ medium</td><td>__ heavy</td><td></td></tr>
<tr><td>Vibration</td><td>__ none</td><td>__ low</td><td>__ heavy</td><td></td></tr>
</table>

Sitting/Standing

Stand constantly ?	__ no	__ yes	__anti-fatigue mats
			__cushioned insoles
Stand on hard surface?	__ no	__ yes	__foot rest
			__better stool/chair
Inadequate chair	__ n o	__ yes	__lean stand
			__sit/stand workstation
Stay in one position constantly?	__ no	__ yes	

Heights and Reaches

Long reaches?
      __ OK __ short employees __ all employees

__reduce surface size
__tilt the work surface
__provide cut-outs

Work done at uncomfortable heights?
      __ OK __ some employees __ all employees

__use lazy Susan
__adjustable stand
__removable stands

Work surface heights in poor
relationship with each other?   __ no __ yes

__chutes and hoppers
__smaller lot size

Clearance

Employees bump into things?

head	__ no	__ yes	__adjustable chair
arms	__ no	__ yes	__adjustable height
body	__ no	__ yes	__adjustable platform
knees	__ no	__ yes	__spring load bins
feet	__ no	__ yes	__align work heights

Employees <u>lean</u> against edge or
uncomfortable equipment surface? __ no __ yes

__pad edges
__round edges
__eliminate obstacles

Ergonomics Issues		Ideas for Improvement or Comments

Tool Design

Grip excessively large or small?	__ no __ yes	__dampen vibration
		__smaller tool
Sharp edges or pressure points on grip?	__ no __ yes	__larger tool
		__modify tool handle
Shock or vibration when used?	__ no __ yes	__use torque bar
		__counter balance tool
Is tool heavy to hold?	__ no __ yes	__lighter tool
		__fixture tool
Designed for right-handers only?	__ no __ yes	__ambidextrous tools

Work Methods

Are there significant differences in the way various employees do the job?	__ no __ yes	__videotape employees
		__employee meetings
If yes, is one method better?	__ no __ yes	__standardize layout
		__motion analysis study
If yes do any employees need additional training to help improve smooth work methods?	__ no __ yes	__employee training

(Answering these questions may take significant study.)

Environmental

Poor lighting; shadows	__ no __ yes	__ventilation diffusers
		__glare guards
Glare; too bright	__ no __ yes	__task lighting
		__replace lights
Floor needs repair / unusually slippery	__ no __ yes	__housekeeping
		__floor maintenance
Temperature extremes; drafty, etc.	__ no __ yes	

Red Flags

Equipment not functioning smoothly	__ no __ yes	__maintenance
		__balance work load
Product piled up	__ no __ yes	__balance work flow
		__poor quality parts
Double-handling	__ no __ yes	
Employee quick fixes (tape, cardboard, rags, etc.)	__ no __ yes	

Administrative Changes

__Job enlargement

__Job rotation

__More frequent, short rest breaks

__Exercise breaks

__Reduced workload

Employee Discomfort

Where do employees typically experience the most discomfort, if any?

hands/ wrists	__ none	__ low	__ medium	__ high
elbows/forearms	__ none	__ low	__ medium	__ high
shoulders	__ none	__ low	__ medium	__ high
neck	__ none	__ low	__ medium	__ high
back	__ none	__ low	__ medium	__ high
legs	__ none	__ low	__ medium	__ high
feet	__ none	__ low	__ medium	__ high

Do you observe anyone . . .

rubbing their elbow or shoulder?
"shaking out" their hands?
looking as thought they are in discomfort?
having red marks, blister, or welts from contact stress?

Employee and Supervisor Input

What are the employee's or supervisor's impressions of what the issues are?

What are their ideas for improvement?

Quality and Efficiency

Do any of the ergonomic issues reduce the ability of the employee(s) to do their jobs efficiently and correctly? (wasted time, errors, scrapped product, needless work, lower yields, extra costs, etc.?) __ no __ yes
(Answering these questions may take significant study.)

Overall

Have you videotaped the task? __ Yes __ No

At this point, it is often best to adjourn to a conference room, watch the videotape multiple times as you complete the rest of the worksheet.

Based on this evaluation, what are your overall conclusions?

Compared with other jobs in this facility, the ergonomic risk factors are: __ low __ medium __ high

What are the top 3–5 issues for this job?   Have you identified any quick fixes?

Example   *Example*
High grasping force to squeeze pliers   Provide floor mat

_____   _____

_____   _____

_____   _____

_____   _____

Use the attached sheets to help brainstorm possible improvements. Make as many copies as needed.

Issue _____

Options for Improvement
(List as many as you can. Include "harebrained" ideas to stimulate creativity.)

_____   _____

_____   _____

_____   _____

_____   _____

_____   _____

Action Items:

Short Term   **Long Term**

_____   _____

_____   _____

_____   _____

Sketch

What can you "trystorm" right now?
(that is, what can you mock up to test an idea)

Summary — Action Items

Based on your findings what are the action items from this evaluation?

Action items can include both:

> *1. recommendations to make specific changes, and*
> *2. assignments to investigate an issue further.*

This is a running list. You can keep adding the next action item until the project is completed.

<u>**Action Items**</u>	<u>**Person Responsible**</u>	<u>**Completion Date**</u>
_____	_____	_____
_____	_____	_____
_____	_____	_____
_____	_____	_____
_____	_____	_____
_____	_____	_____
_____	_____	_____

Supplementary Tools

There are dozens of additional analytic tools standard to engineering that can support the ergonomics problem-solving process.

One example is to formally list the cause of each ergonomics problem. Sometimes the situation is so self-evident that this extra step is nonproductive. Other times, it is valuable to keep you focused on what is actually the problem.

Ergonomics Issues	Causes	Options for Improvement
Long reach; bent back	Parts containers on floor	• Place on stand • Tilt stand • Use smaller containers

Root cause analysis, fishbone analysis, and any number of other tools can be useful.

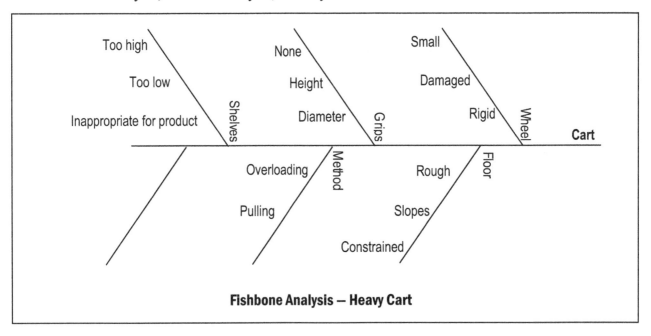

Fishbone Analysis — Heavy Cart

Ergonomics Self-Evaluation Worksheet

Take a close look at each step of the task you are doing. Are there ways to modify the workstation, tools, or equipment to make it better for you? Ask yourself:

1. Do any specific aspects of the job cause discomfort or frustration?

2. Can I make any adjustments or changes by myself with the equipment at hand?

3. Do I have any ideas that could be passed on to a Team Leader or Engineer?

Department _____ Date _____

Task _____ Shift 1st _____ 2nd _____ 3rd

Your name _____

Not all ideas will be feasible,
but all ideas will be given consideration.

Ergonomic Issue	Things I Can Do	Ideas for Improvements
1. Awkward postures?		
bent wrists		
elbows from body		
bent/twisted back		
bent neck		
2. Excessive forces?		
grasping or pinching forces		
push/pull arm forces		
loads on back		
3. Excessive motions?		
hands		
arms		
back		
4. Anything not in easy reach?		
reach envelope		
5. Not at right height?		
elbow height		
height relationships		

Ergonomic Issue	Things I Can Do	Ideas for Improvements

6. Unnecessarily fatiguing?

exhausting

static loads:
- grip
- arm

7. Pressure points?

tool grip

hard edges/surfaces

hard floor

8. Poor clearance and access?

bump/not fit

can't see

9. Freedom to move & stretch?

Constant sitting

Stand in one place

10. Uncomfortable environment?

vibration

temperature extremes

glare, shadows, too bright or dark

Innovation Potential

(Provide sketches)

What improved types of tools are possible?

What type of mechanical assists might be used?

What changes in layout would help?

Would a sit/stand workstation be appropriate?

Would it be feasible to provide a:
__ stool   __ lean stand   __ footrest

Is there a completely different way of doing the job?

Walkthru Survey — MSD Risk Factors

0 = Seems OK,   1 = Medium,   2 = High

Area and job or task	Back			Arms			Wrists			Bent Neck	Shock Vibra-tion	Total
	Bent/twisted	Load	Motions	Reach	Load	Motions	Bent/twisted	Load	Motions			

(Electronic version is formatted in landscape orientation.)

Ergonomics Checklist — Computer Workstation

Name _____   Report to Whom _____

Dept. _____   Phone #   _____

Use computer more than 4 hours daily?___   Use phone more than 1 hour daily?___

Any employee discomfort?   Low back? ___   Neck? ___   Shoulders? ___   Wrists/hands?___

Issue	Cause	Options
POSTURE **Bent or twisted neck**	__ Monitor not in front __ Monitor too high __ Monitor too low __ Cradle telephone in neck __ Paper lies flat on desk __ Copyholder not in best place	__ Move in line; change layout __ Put monitor on swing arm __ Lower, e.g., move hard drive to floor __ Raise: books, risers, stand, arm, etc. __ Headset __ Copyholder (__ extra sturdy) __ Place adjacent to monitor
Rounded, unsupported, or twisted back	__ Insufficient lumbar support __ Sit too far forward in chair __ Monitor or keyboard not in front	__ Adjust height of seat back __ Lumbar support cushion __ Better chair __ Change layout __ Better workstation
Bent wrists	__ Keyboard tray not at right height __ Keyboard too thick __ Traditional straight keyboard __ Keyboard not at correct tilt relative to forearm __ Mouse too large __ Inappropriate design of mouse	__ Raise or lower __ Provide wrist rest __ Contoured natural keyboard __ Adjust feet on keyboard __ Adjust keyboard tray __ Provide adjustable keyboard tray __ Adjust chair height __ Adjust keyboard height __ Mouse pad wrist rest __ Smaller mouse __ Improved-design mouse __ Switch to touch pad/trackball
Shoulders hunched; **Elbows away from sides**	__ Desk/keyboard too high __ Insufficient arm support __ Reach for mouse __ Reach for phone, etc.	__ Lower desk or keyboard __ Provide keyboard tray or arm __ Raise chair __ Better chair to permit adjustment __ Adjust arm rests __ Add cushions to arm rests __ Add armrests __ Put mouse on keyboard tray __ Provide keyboard tray with space for mouse __ Move closer, use headset

Issue	*Cause*	*Options*
STATIC LOAD	__ Constantly sit	__ Break up tasks; do alternate work
		__ Take regular, short stretch breaks
		__ Adjustable sit/stand workstation
		__ Keyboard & monitor arms that raise high enough to permit standing
		__ Adjust chair occasionally
	__ Unsupported arms	__ Adjustable arm rests
		__ Better chair to permit adjustment
	__ Discomfort from constantly holding mouse	__ Alternate hands
		__ Switch to touch pad/track ball
PRESSURE POINTS		
Behind knees	__ Chair too high	__ Lower chair
		__ Provide footrest
		__ Adjust footrest
Buttocks	__ Chair too low	__ Raise chair (__ and raise desk)
Forearms	__ Lean against desk edge	__ Adjust worksurface height
		__ Pad edge or round down edge
CLEARANCE		
Thighs	__ Desk drawers	__ Remove drawers
		__ Replace with thin pencil drawer
	__ Keyboard tray or support arm	__ Replace with thin tray or arm
		__ Raise desk
Knees	__ File drawers (traditional desk)	__ Remove drawers
		__ Replace desk with "L" workstation
		__ Move items from under desk
LIGHTING	__ Glare	__ Remove bulbs
		__ Change diffusers
	__ Too bright	__ Add dimmers
		__ Provide task light
	__ Shadows	__ Provide indirect lighting
		__ Provide glare screen
		__ Close window coverings
		__ Position monitor perpendicular to windows
NOISE	__ Printers or copying machines	__ Isolate noisy equipment
	__ Telephone ringing	__ Reduce ring level
		__ Use alternate sounds
	__ Other nearby conversations	__ Change area layouts to isolate voices
		__ Reduce distracting conversations
GENERAL		__ Provide training on adjustments, layouts, and key concepts

Evaluator(s) _____

Date: _____

Employee Feedback Forms

One the following pages you will find a set of forms that you can use to obtain feedback from employees.

The first survey form is for discomfort. There are some pros and cons regarding the use of these forms. Consequently, the sample form is followed by an overview of the issues and some guidance.

Subsequent to the discomfort survey are several additional forms that may be helpful in obtaining feedback on a variety of topics. These should be self-descriptive. It is likely that you will want to customize these forms to fit your situation.

Employee Discomfort Surveys

Employee discomfort can be measured thru the use of self-administered questionnaires. Typically, employees are asked to rate on a scale of one to five any discomfort they might experience on the job. The questionnaires are usually simple one- or two-page forms that are administered anonymously. Results can be summarized, statistically analyzed, graphed, and evaluated. The purpose of this type of survey is to gain an overview of a particular department, operation, or area.

Two examples of final results are shown below. The first relates to a department of 22 employees where the surveys were administered about six months apart to evaluate the effectiveness of ergonomic changes. The second was administered to a smaller group of employees every 45 minutes, again to evaluate the effectiveness of improvements.

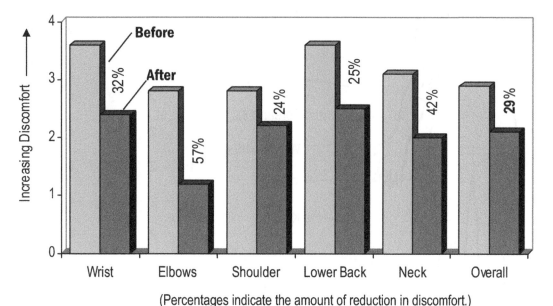

(Percentages indicate the amount of reduction in discomfort.)

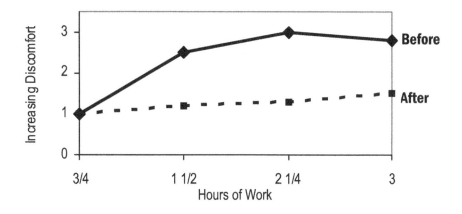

Ergonomics "Body Check"

The following is a survey of how your body feels as a result of your job. Please complete the following to the best of your abilities. Obviously, pain and discomfort can be caused by household chores, sporting, and leisure activities; but the concern here is for any work-related problems.

Department _____   Job _____

Do you experience discomfort or pain in any part of your body as a result of your day-to-day work activities. For those body parts affected, please circle the score which you feel best describes your level of comfort.

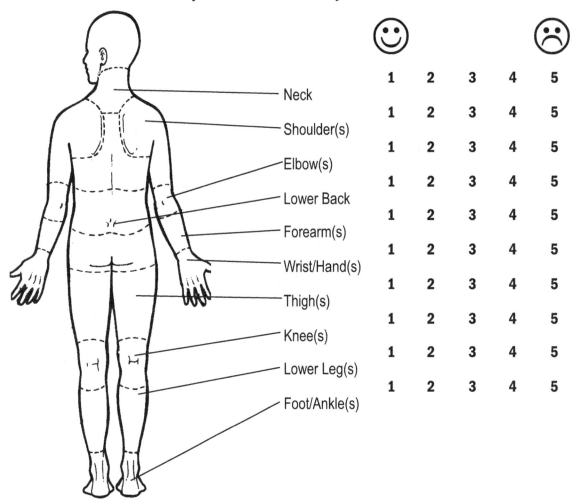

	☺				☹
Neck	1	2	3	4	5
Shoulder(s)	1	2	3	4	5
Elbow(s)	1	2	3	4	5
Lower Back	1	2	3	4	5
Forearm(s)	1	2	3	4	5
Wrist/Hand(s)	1	2	3	4	5
Thigh(s)	1	2	3	4	5
Knee(s)	1	2	3	4	5
Lower Leg(s)	1	2	3	4	5
Foot/Ankle(s)	1	2	3	4	5

Is there a specific aspect of the job that you feel causes discomfort for you?

Additional comments or suggestions?

Uses of discomfort surveys

Targeted areas — The best use of discomfort surveys is probably in a single work area that you have chosen to focus on and is slated for improvements. Results can provide useful insights on potential problems, or at the very least, can confirm what the ergonomist or TeamErgo members have observed. The same kind of information can also be obtained from one-on-one discussions with employees in the area; but this type of survey has its special value, particularly in generating numbers and graphs.

Facility-wide — It is also possible to survey an entire workplace with the objective of helping to prioritize areas or obtain baseline information. However, such use typically entails a number of potential cautions.

The most important concern is that expectations can be raised beyond the ability of your organization to respond. That is why these surveys work better in targeted areas — you already know that you will be making changes.

Also, this type of survey is obviously subjective, and it is possible for results to be skewed because of factors unrelated to ergonomics. For example, employees may not understand the purpose of the survey, and may inflate or understate responses for any number of reasons. Again, targeted areas provide the opportunity to be more precise and understandable regarding why you want the information. Experience is that the survey results tend to be more accurate because the employees reflect more on the questions with the expectation that something will be done about any problems.

In either situation, however, when interpreted appropriately and administered as part of an active ergonomics program, a discomfort survey can provide helpful insights on problem tasks. It is a tool, and like all tools, has its special uses, and may or may not be appropriate in any given situation.

Terminology

Many employers are uncomfortable with using the term "discomfort," since it presumes that there is a problem. Consequently, the terminology adopted in these situations tends to be "comfort survey."

Other employers respond that the reason they are using the survey is because there are indeed problems in targeted areas so they should call things for what they are. Similarly, professional ergonomists tend to use "discomfort survey" when speaking of the generic method in any setting other than in front of employees.

That convention is used here. In this text, the generic method is called a discomfort survey. In the survey form provided with this kit, the term printed on the form itself as distributed to employees is a very neutral "Body Check."

Another Pandora's box?

Often there is a concern that administering a discomfort survey can create problems. However, it has been my experience in all cases that nothing of the sort has occurred. If the purpose of the survey is make task improvements, and it is administered in an atmosphere of trust and sincerity, the experience should be positive.

Tips

1. *The first rule of employee surveys (ergonomics or otherwise) is, "Do not administer the survey if you are not prepared to respond to the results."*
2. You may wish to adapt the labels "department" and "job" that are adjacent to the blanks to fit the situation you are surveying.
3. When a number of work areas are involved, it is a good idea to provide codes for departments and jobs (or whatever areas you are surveying). Otherwise it is too easy for employees to use different names for the same thing, and thus confuse the results.
4. It is helpful to administer the survey forms at the conclusion of an employee training session, so that the purpose of the survey and its objectives can be understood more clearly (and so that all the forms can be collected as the participants leave the room).

Comparison with proactive medical management

There are many similarities between the simple discomfort surveys as described above and the more complex "proactive medical management." Specifically, both seek to evaluate pain in different parts of the body and both might use exactly the same form. However there are crucial differences:

Anonymity — A crucial difference is that a discomfort survey is anonymous, while proactive medical management requires the name of the employee.

Objectives — The difference in obtaining individual names or not stems from the objectives. A discomfort survey seeks to learn about the *jobs* and what activities might be causing problems among employees. The purpose of proactive medical management is to identify *employees* at earlier stages of MSDs in order to permit speedier treatment. To be sure, the resulting data can be studied to learn about certain work areas, but that is not the primary purpose.

Administration — A discomfort survey is self-administered, usually as part of a training class on ergonomics. For medical management, a nurse usually administers the questionnaires in conjunction with a hands-on physical. Also, the nurse might seek more information about the type of discomfort experienced (sharp pain, throbbing pain, etc.).

Occurrence — Many companies routinely administer discomfort surveys, but proactive medical management is not a common practice at the time of this writing.

Employee Feedback Form
Ergonomics Committee

The results will be used only by the Ergonomics Committee to improve our effectiveness.

{Sample: You will need to modify this to fit your workplace.}

Circle the appropriate answer:

1. Years employed here	Less than 1	1–5	5–10
	10–20	20–30	More than 30
2. Job	Driver	Stocker	Supervisor
	Assembler	Shipper	Other
3. Where do you MOSTLY work?	Dock	Racks	Assembly
	Dept A	Dept B	Dept C
	Maintenance	Building Services	
4. Shift	Day	Afternoon	Night

	Yes	No
5. I know at least one member of the Ergonomics Committee.	Yes	No
6. I have reviewed the Job Safety Analyses (JSAs) for my job.	Yes	No
7. I have been trained in Standard Operating Procedures (SOPs).	Yes	No
8. I know what *ergonomics* means.	Yes	No
9. I have seen ergonomic improvements in my area.	Yes	No
10. I know where to go (or who to talk to) to make suggestions concerning ergonomics.	Yes	No
11. I know what kinds of things can cause a musculoskeletal disorder.	Yes	No
12. I know where to go (or who to talk to) if I were experiencing any work-related pain (that is, beyond normal aches and pains).	Yes	No

Indicate how much you agree with the following statements:

	Strongly Disagree	Disagree	Don't Know	Agree	Strongly Agree
13. The Ergonomics Committee addresses problems that concern me.	1	2	3	4	5
14. The monthly safety talks provide valuable information to me.	1	2	3	4	5
15. The company cares about ergonomics.	1	2	3	4	5

Think about the PHYSICAL REQUIREMENTS OF YOUR JOB (that is, LIFTING, BENDING, TWISTING, REPETITIVE ARM AND HAND MOTIONS, etc).

	Strongly Disagree	Disagree	Don't Know	Agree	Strongly Agree
16. I am satisfied with my physical work requirements.	1	2	3	4	5
17. Physical work requirements are getting better.	1	2	3	4	5

	Never	Rarely	Occasionally	Often	Always
18. I experience pain or discomfort (beyond everyday aches and pains) from my job.	1	2	3	4	5

Ergonomics Suggestion Form

Your ergonomics team appreciates any suggestions you have for making improvements. We cannot promise we will implement every idea, but we will promise to consider every idea.

Name* _____

Work area _____

Please describe the problem:

Do you have any suggestions for making improvements?

Do you want a member of the Ergonomics Committee to talk with you?

*Your name is only required if you want feedback or someone to talk to you.

New Equipment Feedback   <u>Pallet Lift</u>

How has the <u>*lift*</u> affected your <u>*back*</u> ?

1	2	3	4	5
Feel considerably worse	Feel a bit worse	No difference	Feel a bit better	Feel considerable better

Does the <u>*lift*</u> make your job easier or harder to perform in a timely manner?

1	2	3	4	5
Considerably *harder*	A bit *harder*	No difference	A bit *better*	Considerably *better*

Should the Ergonomics Committee continue to try to find possible changes like this for other employees?

1	2	3	4	5
Not worth the time and money	Probably not	I don't care	Yes, probably	Absolutely yes

Which brand do you prefer?

       Yellow one ___

       Red one ___

Why?

Other comments?

Chapter 11
Training Slides and Script

The following section provides a sample script for providing training to the people in your organization, along with some tips for preparation. The script consists of both suggested points to make and reproductions of slides that are in the accompanying electronic Microsoft PowerPoint® presentation. As mentioned previously, you may wish to take pictures of your own from your facility and insert them into the slide program.

There are also several commercially available training videotapes that cover roughly the same information (including one by the author using this framework and easily supplemented with video footage from your site). You may wish to use one of these videotapes in combination with, or instead of, this PowerPoint presentation. The advantage of packaged videotapes is that they explain many items that you may not feel comfortable with, or in those cases where you need to conduct multitudes of sessions, the videos save some wear and tear on your throat. Alternatively, you can use videos as refreshers in the future.

In all cases, you will most likely want to make at least some part of the presentation to address items that are specific to your facility. Thus, at least parts of the following script should help you in any event.

Electronic Handouts

The downloadable materials that come with this kit also contain short handouts in Microsoft Word. You may reproduce copies of these (for your facility only!) to use as handouts. Additionally, there are many commercially available booklets that you can use, many of which are packaged with the videotapes mentioned above, including the author's. Finally, you may want to prepare and distribute the following:

- A list of TeamErgo members;
- An outline of your site's ergonomics program, including information on how to request needed improvements and procedures for seeking a medical evaluation.

Length

This presentation is designed for a session that lasts about one hour with discussion, but you can adapt it for different lengths. Example discussion questions are found at the end of the script.

Planning the Training:

This list of questions should help you plan training, whether you are the on-site person responsible for the training or a corporate coordinator helping ensure that the training is properly planned.

1. Who will do it?

Who would you expect to conduct the employee sessions: (a) coordinator/committee members, (b) supervisors, (c) facility trainers, or (d) professional ergonomists?

2. Who's trained first?

With which group of employees would you start the training? Is there a priority or pilot group?

3. How does it fit in?

What other types of training have been conducted at this location? What would you do to help ensure that the ergonomics training fits into this other training?

4. How do you encourage early reporting?

Encouraging employees to report symptoms early is a key part of an MSD prevention program. What is the specific procedure that you need to tell them (see their supervisor, go to the medical department, get permission to go to the clinic, etc.)? Is there an on-site nurse who can participate in this part of the program?

5. "You're going to do what?"

What are good responses to concerns about encouraging employees to report symptoms early? ("It will open a Pandora's Box." "Telling people symptoms creates the symptoms.")

6. Who keeps the notes?

If you use these sessions to generate lists of ergonomics issues in the employees' work areas and obtain ideas for improvement, who would be responsible for recording these comments and following thru?

7. How do you respond?

How would you handle a discussion where participants start ventilating considerable complaints or describe serious problems?

8. What's their role in the program?

How would you customize the What-You-Can-Do portion of the program to specifically fit your workplace or company?

9. Who will customize the training?

Who is responsible for preparing any customized portions of the presentation and the handouts of committee members and the company's program?

10. How was it?

How can you evaluate if the training was effective?

Your Preparation

Go thru the slides and script. Then go thru the Power-Point presentation reviewing the bullets that go with each slide. Go thru the presentation several times until you know what you are going to say as each slide appears.

Instructor Tips and Options

1. Make your objectives for the session clear to the participants.

2. Review the agenda so that everyone knows the direction in which you're going.

3. Use specific examples from your workplace to bring home the general points made in the handouts, videos, or slides.

4. Use the word "you" often to involve participants. Talk with people, rather than lecture.

5. Use an interactive discussion style and try asking open-ended questions about ergonomics issues in the participants' work areas.

6. Encourage discussion, but watch for signs that people are getting off the topic. Sometimes you may need to cut off a lively interaction with a comment like, "These are all good points, but we have more to cover, so we must move on."

7. If you don't know the answer to someone's question, say that you don't know. You are (probably) not an expert in the field, and no one can be expected to know everything.

8. Typically, you are trying to generate enthusiasm for ergonomic improvements, but avoid raising expectations too high. It can be a delicate balance to maintain. For example, when discussing chairs, you may need to state that you are not indicating that everyone will get a new chair, but that they are being phased in (or whatever your situation is). In all cases, avoid making promises that you are in no position to keep.

9. Humor can be very effective as part of a presentation, but make sure it is not at the expense of anyone, whether anyone present or any class of individuals.

10. Provide a five-minute break every hour, and if feasible, have a wellness instructor lead the group in stretching exercises.

11. Dim the lights for showing slides and overheads, but do not turn them off completely, so you can see the participants and they can see you (and their training handouts). Make sure you know where the light switches are beforehand.

12. Run through your presentation very quickly prior to starting so that you can remember exactly what you are planning to say.

The Best Preparation You Can Get ...

...is going around in your facility, talking with people, and getting a good feel for what your specific issues are and what it will take to make needed improvements. Practice doing some workstation evaluations and thinking of ways to make improvements. Coupled with all the information in this kit, you should be well prepared to conduct good basic training session.

General Agenda

The general agenda for awareness sessions for either managers or employees is the same. The only difference is the level of detail provided and the framework for addressing personal responsibility.

- What is ergonomics?

- What are musculoskeletal disorders?

- Ergonomic issues in your work area.

- This facility's ergonomics program.

    - TeamErgo: structure and members.
    - Activities for evaluating and improving jobs.
    - Medical issues and procedures for getting help.

- What you should do.

- Option: Employee survey.

About the PowerPoint Presentation

The PowerPoint presentation that is part of this kit provides you with a basic set of slides that you may use. You can easily show them directly using a large computer monitor or an LCD projector, or you can convert them into overhead transparencies. Simple handouts generated in Microsoft Word are also included.

The fonts used for the slides and handouts are Arial Narrow Bold and New Times Roman, which should be standard fonts on most pieces of equipment, and thus ready to use as is. Obviously, you may customize these to fit your own preference.

The slide design scheme is a fairly plain layout. Again, the software makes it fabulously easy to add color and design features of your choice to the best of your capabilities.

You may wish to make some slides of your own:

- injury cases and types at your facility
- workers' compensation costs at your facility
- overview of your ergonomics program
- overview of procedures, such as how to report problems

You can customize the program to emphasize issues that are particularly important to your facility and delete ones that are not. If you want to really increase the value of the presentation, you may want to take some photos of your own workstations to use (either using a digital camera or by scanning prints or 35mm slides). In this case, you can simply use the generic ones as guides and take a picture to represent each of the issues. You may also increase the value of your presentation by showing videos of various work areas from your facility. The more that people see of your own site, the more they are likely to connect with the training.

In the following chapter, each of the slides is reproduced along with a general script. These materials should enable you to conduct a basic session. Obviously, you will need to change some parts of the script that deal with aspects of your particular facility.

Good luck and have fun.

This page itemizes a list of objectives on which the sample presentation was based. There is an electronic version of this page available to enable you to customize it as necessary for your own needs or style of writing training objectives.

Training Program Objectives

All participants will ...

1. Gain a general understanding of ergonomics.

2. Learn to identify basic ergonomics issues in their work areas and some ideas for improvement.

3. Learn the basics for adjusting equipment and optimizing layouts in their own workstations.

4. Understand the employer's ergonomics program.

5. Understand their responsibilities and role in the process.

6. Understand how to work effectively with the facility's ergonomics committee or task force.

   - How to recognize issues
   - Where to go and how to deal with problems

Options:

7. Use this session to identify ergonomics issues and possible improvements.

8. Fill out the employee survey.

You may wish to customize and/or make a handout of this page.

In this column are reproductions of the slides that are in the PowerPoint presentation.

This column provides a sample script, which is summarized in bullet format in the PowerPoint file (in "notes page" view) and which you probably should print out.

You can simply read the script if necessary, but it is obviously better if you learn to speak in your own words based on following the bullets. You will also need to adapt certain phrases to fit your facility.

Ergonomics

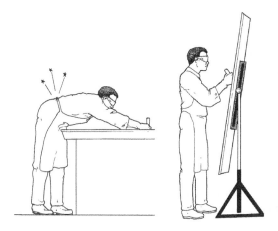

This program focuses on providing a basic awareness of ergonomics issues at work.

We all want to make sure that no one is working in any pain or discomfort related to their tools, equipment, or tasks. We want to minimize awkward and uncomfortable positions that you might be working in.

The following information is designed to provide a basic overview of issues so that you can evaluate your own work areas, make feasible changes, or request improved equipment. We will also cover what you should do if you are having a problem.

Agenda

- **What is ergonomics?**

- **What are Musculoskeletal Disorders?**

- **What are applications in the workplace?**

- **What are activities at this site?**

- **What you can do.**

The topics included here are:

A very basic description of the field of ergonomics and why we need to pay attention to the message.

A quick overview of Musculoskeletal Disorders — MSDs — which amount to discomfort, and wear and tear that can occur from thousands of activities, including those at work.

We want to address how this all fits into the workplace, what is occurring here at this site, your responsibilities, and the steps you can take to improve your own well-being.

What Is Ergonomics?

There are many good ways of defining ergonomics:

A good slogan that sums up the field is "fit the task to the person, not the person to the task."

The term *ergonomics* was coined from the Greek words *ergon* (meaning "work") and *nomos* (meaning "rules); so the literal meaning is "the rules of work." And everyone needs to play by the rules of work.

Another way to understand the field is that it provides a method for finding ways to work smarter, and not harder.

Ergonomics focuses on how tools, equipment, layout, and overall organization of work affect your well being and your ability to get your job done easily. It's all about making things user-friendly.

Applications for Ergonomics

Manufacturing   **Home & Leisure**

Service   **Office**

The possible applications for ergonomics include about every activity that people do both on and off the job.

- Manufacturing
- Service industry
- Office work
- Home chores
- Leisure activities

Ergonomics provides a way of thinking about how to do all types of tasks to make them easier and more efficient. Ergonomic improvements can enhance your ability to do almost any task more safely and effectively. In these materials, however, we are going to focus on the workplace.

Safety, Quality, and Efficiency

The goals of ergonomics are to improve employee well-being, as well as to enable people to do their jobs better.

For example, look at this person. Does this look like a user-friendly set-up? Do you think she might be sore at the end of the day? Is she in a position to do her job right the first time working like this? Can she work efficiently?

This is what we mean by trying to fit the task to the person, and the reasons for doing this are to promote safety, quality, and efficiency.

Principles of Ergonomics

The word "ergonomics" is unusual, and it sounds like it ought to be difficult. But it is not necessarily so. On one level, ergonomics can be summarized in a series of principles.

We will go thru these principles one by one so you can understand the concepts. You will very readily see that you can apply these principles at work, at home, or any other place you happen to be.

Principle 1
Work in Neutral Postures

Your posture provides a good starting point for evaluating the tasks that you do. The best positions in which to work are those that keep the body "in neutral." We will go thru what that means for (1) your back and neck, (2) your arms, and (3) your wrists.

Maintain the "S-Curve" of the Spine

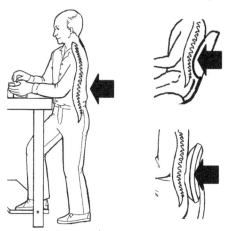

Your spinal column is shaped more or less like an "S." It is important to maintain the natural S-curve of the back, whether sitting or standing. The most important part of this "S" is in the lower back, which means that it is good to keep a slight "sway back,"

Additional lumbar support is often helpful to maintain the curve in the small of your back.

This is obviously not the S-curve. It's more like a C-curve, which can be very stressful on your spine.

Something like this could be improved by changing worksurface heights and reaches (more about this later).

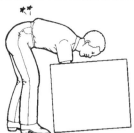

This is even worse. This is an inverted "V."

[Incidentally, we've always heard the phrase, "lift with your legs, not with your back." Can a person in a setup like this lift with the legs? Of course not — the container sides are in the way. That's why engineering changes are good.]

Making this task better involves getting the load up off the floor. One way of doing this is by using a scissors lift table.

[There are many versions of lifts like this on the market. Part of what we need to do is find which ones are best for our needs.]

Another way is by using a tilter like this. Or there may be more ways of making improvements.

By the way, for any of the examples I'm showing, I'm not necessarily saying that it is always feasible to install this type of equipment. The point is to get each of us thinking about where there are potential problems and what could be done.

Finally, it is also stressful to twist your back like this. One thing we want to look for is any location where we can eliminate twisting like this, maybe by rearranging the work area layout.

The Neck

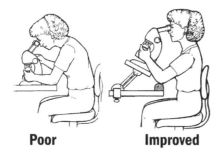

Poor   **Improved**

The neck bones are obviously part of the spinal column and thus are subject to the same requirements of maintaining the S-curve. Prolonged twisted and bent postures of the neck can be as stressful as for the lower back.

The best way to make changes is usually to adjust equipment so that your neck is in its neutral posture.

Or as another example, we've all probably gotten sore necks from crooking the telephone in our necks like this. In this particular instance, a headset would solve the problem.

More generally, there are thousands of ways to improve neck posture, depending upon what is causing the neck to be bent.

Keeps Elbows in and Shoulders Relaxed

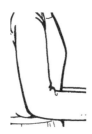

The neutral posture for your arms is to keep you elbows at your sides and your shoulders relaxed.

This is pretty obvious once you think about it, but we don't always do it.

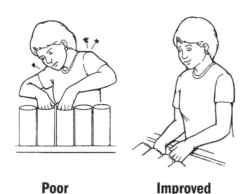

Poor   **Improved**

Here's an example of changing a workstation to get the arms in neutral. In the illustration at the left, the product is too high, and the employee is hunching her shoulders and winging out her elbows.

In the right-hand illustration, the product is lowered and the shoulders and elbows drop to their relaxed position.

Keep Wrists in Neutral

There are several good ways to think about wrist posture. One way is to keep the hand in the same plane as the forearm, as this person is doing here by using a wrist rest along with the computer mouse.

A slightly more accurate approach is to keep your hands more or less like they would be when you hold the steering wheel of your car at the 10 and 2 o'clock position — slightly in and slightly forward.

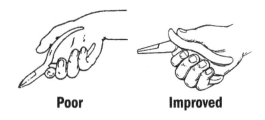

Poor　　　**Improved**

Here's an example of how this principle applies to tool design. Working continuously with the pliers as shown in the left-hand picture can create a lot of stress on the wrist. By using pliers with an angled grip, however, the wrist stays in its neutral posture.

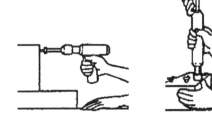

Always be aware that situations change depending upon the task. Note here that two differently designed tools are used depending upon the application.

The key is the wrist posture. Look at the wrist to determine the right tool.

Summary — Neutral Postures

In summary, the best positions in which to work are those that keep the body "in neutral." This means:

- The back with its natural "S-curve" intact
- The elbows held naturally at the sides of the body
- The wrists in neutral position

These rules apply to whether you work on a computer or any other type of work. (Of course, this doesn't mean never, ever do anything different. It applies only to sustained work.)

**Principle 2
Reduce Excessive Force**

Excessive force on your joints can create a potential for fatigue and injury. Consequently, Principle 2 of ergonomics is to identify specific instances of excessive force and think of ways to make improvements.

For example, pulling a heavy cart might create excessive force for your back. To make improvements it might help to make sure the floor is in good repair, that the wheels on the cart are sufficiently large, and that there are good grips on the cart. Or a power tugger might be needed.

Or another example of reducing force is to use a hoist for lifting heavy objects, like this vacuum hoist in the drawing.

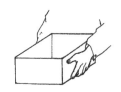

Another kind of example is having handholds on boxes or carrying totes. Having the handhold reduces the exertion your hands need to carry the same amount of weight.

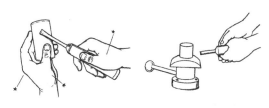

Or as a final example, if you need to hold onto an item, then work on it with your other hand, you end up working against yourself. It's a lot easier if you can put the item in a fixture. Furthermore, sometimes it is then possible to use both hands on a tool, which cuts force even more.

Point:

There are thousands of other examples. The point is to recognize activities that require excessive force, then think of any way you can to reduce that force.

Principle 3
Keep Everything in Easy Reach

The next principle deals with keeping things within easy reach.

Reach Envelope

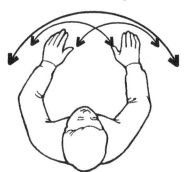

One concept is to think about the "reach envelope." This is the semicircle that your arms make as you reach out. Things that you use frequently should ideally be within the reach envelope of your full arm. Things that you use extremely frequently should be within the reach envelope of your forearms.

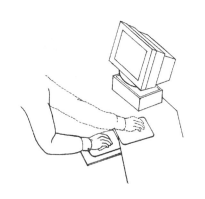

Much of the time, problems with reach are simply matters of rearranging your work area and moving things closer to you. This is not exactly a hard concept to grasp; what is difficult is having the presence of mind to notice and change the location of things that you reach for a lot.

Often it is a matter of habit — you are unaware that you continually reach for something that could be easily moved closer.

Or sometimes, the work surface is just too big, causing you to reach across to get something. One option is just to get a smaller surface. Another option is to make a cutout — this way your reaches are cut, but you still have plenty of space for things.

Or another common problem is reaching into boxes. A good way to fix this is to tilt the box.

Once again, there are thousands of other examples of ways to reduce long reaches. The point is for you to think about when you make long reaches, then figure out how to reduce that reach.

Principle 4
Work at Proper Heights

Working at the right height is also a way to make jobs easier.

Do most work at elbow height

A good rule of thumb is that most work should be done at about elbow height, whether sitting or standing.

A real common example is working with a computer keyboard. But, there are many other types of tasks where the rule applies.

Exceptions to the Rule

There are exceptions to this rule, however. Heavier work is often best done lower than elbow height. Precision work or visually intense work is often best done at heights above the elbow.

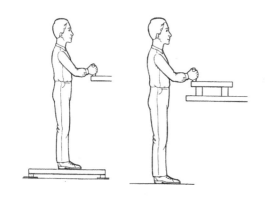

Sometimes you can adjust heights by extending the legs to a work tables or cutting them down. Or you can either put a work platform on top of the table (to raise the work up) or stand on a platform (to raise YOU up).

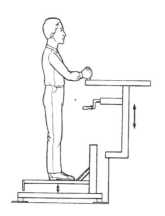

Or to be a little more complicated, there are ways to make stands and work tables instantaneously adjustable with hand cranks or pushbutton controls.

Principle 5
Reduce Excessive Motions

The next principle to think about is the number of motions you make throughout a day, whether with your fingers, your wrists, your arms, or your back.

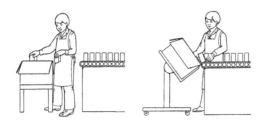

One of the simplest ways to reduce manual repetitions is to use power tools whenever possible.

Another approach is to change layouts of equipment to eliminate motions. In the example here, the box is moved closer and tilted, so that you can slide the products in, rather than having to pick them up each time.

Or sometimes there are uneven surfaces or lips that are in the way. By changing these, you can eliminate motions.

As always, there are more examples, but you should be getting the idea.

Principle 6
Minimize Fatigue and Static Load

Holding the same position for a period of time is known as *static load*. It creates fatigue and discomfort and can interfere with work.

A good example of static load that everyone has experienced is writer's cramp. You do not need to hold onto a pencil very hard, just for long periods. Your muscles tire after a time and begin to hurt.

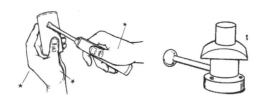

In the workplace, having to hold parts and tools continually is an example of static load. In this case, as we have seen previously, using a fixture eliminates the need to hold onto the part.

Having to hold your arms overhead for a few minutes is another classic example of static load, this time affecting the shoulder muscles. Sometimes you can change the orientation of the work area to prevent this, or sometimes you can add extenders to the tools.

Having to stand for a long time creates a static load on your legs. Simply having a footrest can permit you to reposition your legs and make it easier to stand.

We're going to come back to this point later.

**Principle 7
Minimize Pressure Points**

Another thing to watch out for is excessive pressure points, sometimes called "contact stress."

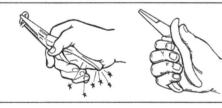

A good example of this is squeezing hard onto a tool, like a pair of pliers. Adding a cushioned grip and contouring the handles to fit your hand makes this problem better.

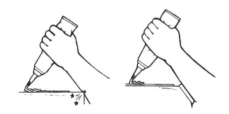

Leaning your forearms against the hard edge of a work table creates a pressure point. Rounding out the edge and padding it usually helps.

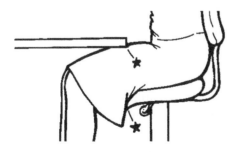

We've all sat on chairs without cushioning and so understand almost everything you need to know about pressure points. A particularly vulnerable spot is behind your knees, which happens if your chair is too high or when you dangle your legs. Another pressure point that can happen when you sit is between your thigh and the bottom of a table.

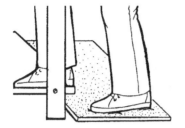

A slightly more subtle kind of pressure point occurs when you stand on a hard surface, like concrete. Your heels and feet can begin to hurt and your whole legs can begin to tire. The answer is anti-fatigue matting or sometimes using special insoles in your shoes.

Like the other basic principles that we've covered so far, pressure points are things that you can look for in your work areas to see if there are ways to make improvements.

Principle 8
Provide Clearance

Having enough clearance is a concept that is easy to grasp.

Work areas need to be set up so that you have sufficient room for your head, your knees, and your feet. You obviously don't want to have to bump into things all the time, or have to work in contorted postures, or reach because there is no space for your knees or feet.

Being able to see is another version of this principle. Equipment should be built and tasks should be set up so that nothing blocks your view.

Principle 9
Move, Exercise, and Stretch

To be healthy the human body needs to be exercised and stretched. You should not conclude after reading all the preceding information about reducing repetition, force, and awkward postures, that you're best off just lying around pushing buttons.

Depending upon the type of work you do, different exercises on the job can be helpful.

- If you have a physically demanding job, you may find it helpful to stretch and warm up before any strenuous activity.

- If you have a sedentary job, you may want to take a quick "energy break" every so often to do a few stretches.

If you sit for long periods, you need to shift postures:

- Adjust the seat up and down throughout the day.

- Move, stretch, and change positions often.

It actually would be ideal if you could alternate between sitting and standing throughout the day. For some tasks, such as customer service, desks are available that move up and down for this purpose (this is not new; Thomas Jefferson built a desk like this for himself).

Principle 10
Maintain a Comfortable Environment

This principle is more or less a catch-all that can mean different things depending upon the nature of the types of operations that you do.

Lighting and Glare

One common problem is lighting.

In the computerized office, lighting has become a big issue, because the highly polished computer screen reflects every stray bit of light around.

But many other types of tasks can be affected by poor lighting, too. Concerns include glare, working in your own shadow, and just plain insufficient light.

One good way to solve lighting problems is by using task lighting; that is, having a small light right at your work that you can orient and adjust to fit your needs.

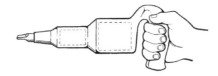

Vibration is another common problem that can benefit from evaluation. As an example, vibrating tools can be dampened.

Summary

So now we have concluded an overview of the basic principles of the field of ergonomics.

Posture	Fatigue and static load
Force	Pressure points
Reach	Clearance
Heights	Stretch and exercise
Motions	Environment

- Work in neutral postures
- Reduce excessive force
- Reduce excessive reaches
- Work at proper heights
- Reduce excessive motions
- Reduce fatigue and static load
- Eliminate pressure points
- Provide for clearance
- Stretch and exercise
- Maintain a good environment

There's a lot more to it than this, but the point here was just to go over the basics. Now we're going to talk about PROCESS, that is, how we're going to start using these principles in our facility.

Putting on Your Ergonomics Glasses

The basics of ergonomics do not need to be hard. Much of it amounts to looking at routine activities from a new perspective — putting on your ergonomics glasses.

Making Improvements

After considering the principles I've mentioned, you are likely to have many ideas for improvements. You may know your job better than anyone else, and may have excellent ideas. That's part of the purpose of this program — to establish a framework for implementing these ideas.

Be aware that not all can be implemented, at least at once. It is helpful to think about improvements in categories of (a) quick fixes and (b) long term.

You may also find that some improvements require a willingness to change habits.

Fix and Adjust What You Can Yourself

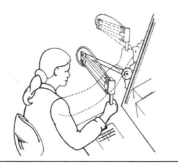

Many of the principles I've mentioned (like keeping things in easy reach) are not exactly difficult concepts to grasp. What is hard is having the presence of mind to think about what you're doing.

It doesn't take special genius to move things closer to you, like this example of learning to adjust an arm that holds a tool to bring it closer to you. It just takes some self-awareness of what you are doing.

Think Before You Work

Think before you work. Can you make any changes to where you put things to make it easier? Less reaches? More comfortable?

Why work the hard way? Wherever you are, think about what you're doing. For example, if you are going to have to lift a load again, why put it on the floor? Put it on a waist-high surface instead.

Make Ergonomics Part of Daily Worklife

The goal is to make this more than a one-shot process. We should make ergonomics part of our everyday worklife.

Musculoskeletal Disorders (MSDs)

One of the important goals of ergonomics is to prevent MSDs, a class of disorders that basically amount to *wear and tear* on the tissue surrounding your joints. Every joint in the body can potentially be affected, but the lower back and upper limbs are the areas of most concern. The term *cumulative* is used to indicate that the trauma *accumulates* over a period of time.

Most of us will experience an MSD of one sort or another in our lives, often sports related, or just lower back pain from everyday life. Most of the time the symptoms are mild and disappear with rest. However, sometimes MSDs can become disabling.

Symptoms

- Soreness, pain, or discomfort
- Numbness or tingling sensations ("pins and needles")
- Weakness and clumsiness
- "Burning" sensations
- Limited range of motion
- Stiffness in joints
- Popping and cracking in the joints
- Redness and swelling

Symptoms of an MSD include: *(read list)*

Now, we're not trying to turn you all into hypochondriacs, but it is important for you to know if you are experiencing anything like this.

Get it checked out.

If you are having problems, get it checked out. Report your problems to your manager and use the established procedure to get a medical evaluation. The earlier a potential medical problem is recognized, the easier it is to treat, and the more likely you can get 100% better.

If caught early, you can treat many MSDs with ice packs and over-the-counter medications (like ibuprofen). If you wait too long, the injury may require surgery.

RISK FACTORS

There are several factors that can increase the risk of MSDs. These can affect the lower back as well as the wrist, elbow, and shoulder. The more factors involved and the greater the exposure to each, the higher the chance of developing a disorder.

Tasks

The following items relate to the activities you do, whether on or off the job.

Repetition

The number of motions made per day by a particular body part.

Force

The exertion required to make these motions.

Awkward Postures

Body positions that deviate from "neutral."

Contact Stress

Direct pressure against any vulnerable part of the body.

Static Load

Using the same muscles for a period of time without change.

Environmental

Exposure to vibrating tools or temperature extremes.

Stress

Certain stressful conditions.

Personal Issues

Certain personal issues can also contribute to MSDs.

Physical Condition

Poor personal fitness can play a role in the development of some MSDs.

Diseases and Conditions

There are also a number of diseases and conditions that can increase the risk of certain types of MSDs, for example, diabetes and pregnancy.

Prevention:

Apply what you learn here

By applying the recommendations from these materials, these risk factors can be reduced and MSDs prevented.

340

WHAT YOU CAN DO	Here are a number of things you can do. These are your responsibilities and what is expected of you.
Request help for any ergonomics problems	Request help for any ergonomics problems from your manager or TeamErgo members, or if you would like an overall evaluation of your work area.
Be willing to try	Be willing to try new work aids and equipment. Often a new item or a changed layout may feel awkward at first, and you need to get used to it.
Know what's adjustable	Know what's adjustable in your work area. Know how to adjust it, and do it routinely — when you start to work, and whenever you need to change.
Get it checked out	If you are experiencing any pain or discomfort that seems beyond the ordinary, get it checked out. Report your problems to your manager and use the established procedure to get a medical evaluation. The earlier a potential medical problem is recognized, the easier it is to treat.
Exercise and keep in shape	Exercise and keep in shape. Take time to stretch, and move around periodically. If you've been working in a certain posture continuously, stretch the opposite way. Strength and flexibility can protect you from some cumulative injuries. General wellness is important.
Think about your off-the-job activities	Our focus here is on work, but think about your off-the-job activities. Your household chores, hobbies and leisure activities may affect your well-being. Apply these principles of ergonomics to your everyday life.
Learn more about ergonomics	We all need to learn more about ergonomics. There are articles being published in newspapers and in many common magazines. TeamErgo and I also have more information if you would like.
Our plan	Let me finish by telling you a bit more about what is happening at this site.

(You will need to flesh this out.)

TeamErgo

One of the things we've formed is an ergonomics team, which consists of ...

(Name names and describe activities.)

Action plans, etc.

What we're planning to do is ...

Your role

What we want you to do is ...

(This can vary — you may just want the participants to be aware of what is happening on site, or you may want to solicit ideas. You need to tell them.}

Problem-Solving Session

If you want to also spend additional time to identify and solve problems, the following are tips and ideas. There are any number of approaches you can take. For example, you can simply have a discussion, or you can go in the participants' work area and do a survey and brainstorm ideas.

Home Videos of Your Site

In either case, it can be helpful to make a videotape of typical issues at your site to help people apply what they are learning and to help get everyone thinking about ideas for improvements. You can show tapes of the participants' jobs and ask for input.

You can also show before-and-after videos or slides of workstations and equipment that you have already improved at your site. Another thing you can do is show a video of how to use adjustable equipment that is common to your operations.

Discussion Points

To generate discussion you can ask these questions:

- Are there activities in your work area that cause discomfort (of the sort described in the slide presentation)?

- Do you have any ideas for improvement (even ideas you've suggested in the past that were never followed up)?

- What equipment is adjustable and how does it work?

- Are some work methods and techniques better than others?

Employee Surveys

The conclusion of the training session is an excellent time to distribute employee feedback forms. You have a captive audience, they know what the survey is all about, they have time to complete it, and they can leave the form in the room as they leave.

Manager and Supervisor Training

Managers and supervisors need to receive the same introduction to ergonomics as do all employees. Additionally, however, these individuals typically need a bit more information and can benefit from a separate presentation.

Much of this type of presentation you will need to prepare yourself, depending upon the specifics at your site. In some cases you may need to discuss a considerable number of issues, and you may need to pull information from other sections of this kit. In other cases, this type of session may be very brief.

Additionally, some organizations have a system of using their managers or supervisors as the trainers. If you are in this category, you will obviously need to spend more time with them to prepare them to conduct training. There is no plan for doing that here, since this type of preparation is highly specific to your organization. However, once again, you should be able to get the background you need from this kit.

With those caveats in mind, the PowerPoint presentation has a number of generic word slides that you may wish to expand upon. If you haven't used this software, don't worry. It is very easy to use and you can quickly create new slides if you need them.

This page itemizes a list of objectives for a presentation to managers and supervisors. Once again, there is an electronic version of this page on the disc to enable you to customize it as necessary for your own needs or style of writing training objectives.

Program Objectives
Supervisors and Managers

All participants should be able to ...

1. Explain to employees in simple terms what ergonomics and Musculoskeletal Disorders are.

2. Identify basic ergonomics issues and ideas for improvement within their work areas.

3. Recognize the situations that may cause MSDs and how these same factors may be contributing to inefficiency or other costs.

4. Understand their responsibilities and role in the program.

5. Understand the company's program in ergonomics.

6. Facilitate, coach, and instruct their employees to make their work areas better.

7. Talk with employees about their jobs and discuss methods for improvement.

8. Understand cost and business implications.

(Option)

9. Be able to conduct a one-hour employee training session.

You may wish to customize sand/or make a handout of this page.

Managers/Supervisors Supplement

Introduction	You've seen the basic program about ergonomics. I hope that it has been useful to you. Now I'd like to highlight a few more items for managers and supervisors.

Benefits for Business	You've learned about ergonomics, and this may have cleared up a lot of the questions you have. You should be able to see that ergonomics isn't about making cushy jobs for people. There are clear benefits for business. Ergonomics can help us make a more efficient workplace as well as making sure no one suffers any disorder.

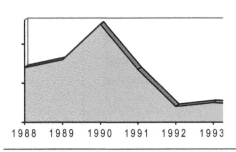

1988　1989　1990　1991　1992　1993

MSDs constitute the major portion of workers' compensation costs in most companies — back injuries, and increasingly problems with wrists, elbows, and shoulders. Ergo programs have prevented these costly injuries and saved money.

You should add as much information as you have on workers' comp costs at your site.

Methods Engineering	Altho our focus is on preventing wear-and-tear injuries, we can also use the same concepts to improve efficiency. In many ways, ergonomics is similar to old-fashioned methods engineering. *You can distribute the handout* 25 Ways Ergonomics Can Save You Money *at this point. You may want to add slides from your site to illustrate these points.*

Your Responsibilities	In the basic program we itemized what we expected of all employees. Here are a number of items that deal with your responsibilities.

Know the Basics of Ergonomics	Know the basic principles of ergonomics. You should be able to explain to people about the field and the general principles as they affect the work environment. TeamErgo and I can always help you get the background you need so that you can be confident about what you know.

Know about MSDs	Know about Musculoskeletal Disorders — general types, symptoms, and causes. You don't need to know a great deal about specifics, but you should be able to explain to people that MSDs basically amount to wear and tear on the human body.
Recognize Issues in Your Area	Recognize the ergonomics issues in your area and activities that have caused MSDs and discomfort. The basic program provided you with most of what you need to know. This item is mostly aimed at getting in the habit of putting on your ergonomics glasses and noticing issues in your area.
Work with Employees to Make Changes	Work with employees to make those changes that you do by yourselves. Some of this is helping people figure out how to rearrange their areas. Some of it is reminding people to change habits (without you becoming the ergonomics police). And some of it is identifying who needs what types of improved tools or equipment. In this last case, you will need to communicate with TeamErgo (or facilities) to obtain needed improvements. *(This is an example of something that you need to customize for your facility.)*
Make Sure People with Problems Get Medical Attention	If an employee reports symptoms, or you see an employee who appears to have symptoms, make sure the employee receives medical attention. Make sure everyone knows the procedures they are supposed to use. One of the issues is spotting people who are having problems, but who are reluctant to report them to you or get medical attention. There are a variety of reasons why this can happen, but you should do your best to promote early recognition of problems. Some things you can watch for in particular are people shaking out their hands (this is caused by the pins and needles that come with wrist problems; it's almost a natural inclination to react as though you were trying to shake water off your hands).

Also watch for people rubbing their wrists, elbows, or shoulders. This might be something else, but it could be related to some discomfort caused by poor ergonomics. Furthermore, it may just make you aware of who is having problems.

Show Concern for Well-Being Solicit Input	Show concern for employees and their well-being. Talk with employees about ergonomics and solicit input and requests. This is a rather self-evident item, but I'm mentioning it to be complete. It's a cost-free way of helping people.
Help Accommodate People with Restrictions	If an employee has work restrictions, help accommodate the employee and see that medical restrictions are observed. Again this is pretty self-evident, but it is part of your responsibility.
Know Policies and Procedures	Be familiar with company policies and programs for ergonomics and Musculoskeletal Disorders. *(Again, you will need to prepare this.)*
Standards and Regulations	*(If you want to cover this, you can make some bullet items from the information that's in this kit.)*
Our Plan	*(You will need to flesh this out.)*
TeamErgo	
Action Plans, etc.	
Temporary Increase in Reported Injuries	One final thing that you should be aware of is that when we start increasing awareness about ergonomics and MSDs, people will naturally report any problems (if they do what we tell them), which will make our recordable injury rate go up. This is OK, since it means we are most likely preventing more serious — and expensive — cases from developing in the future. Ultimately, though, the recordable cases will drop. *(If you have bonuses based on recordable injuries, you most likely would want to change the system to eliminate basing anything on simple recordable cases of MSDs. Otherwise there's an incentive NOT to report problems.)*

Supervisor's Responsibilities

1. Know the basic principles of ergonomics.

2. Understand the role of ergonomics in identifying wasted motions, unnecessary activities, fatiguing conditions, and other sources of inefficiency.

3. Understand the basic problem-solving process.

4. Know about Musculoskeletal Disorders — general types, symptoms, and causes.

5. Recognize the ergonomics issues in your area and activities that have caused Musculoskeletal Disorders.

6. Work with employees to make those changes that you can do by yourselves. Communicate with TeamErgo regarding improvements.

7. If an employee reports symptoms, or you see an employee who appears to have symptoms, make sure the employee receives medical attention.

8. Show concern for employees and their well-being. Talk with employees about ergonomics and solicit input and requests.

9. If an employee has work restrictions, help accommodate the employee and see that medical restrictions are observed.

10. Be familiar with company policies and programs for ergonomics and Musculoskeletal Disorders.

You may wish to customize and/or make a handout of this page.

Index

Printed and bound by CPI Group (UK) Ltd, Croydon, CR0 4YY

23/10/2024

01778250-0019